NASA AT 40: WHAT KIND OF SPACE PROGRAM DOES AMERICA NEED FOR THE 21st CENTURY?

AF817434

HEARING

BEFORE THE

SUBCOMMITTEE ON SPACE AND AERONAUTICS

OF THE

COMMITTEE ON SCIENCE

U.S. HOUSE OF REPRESENTATIVES

ONE HUNDRED FIFTH CONGRESS

SECOND SESSION

OCTOBER 1, 1998

[No. 90]

Printed for the use of the Committee on Science

U.S. GOVERNMENT PRINTING OFFICE

53–067CC WASHINGTON : 1998

For sale by the U.S. Government Printing Office
Superintendent of Documents, Congressional Sales Office, Washington, DC 20402
ISBN 0-16-058025-0

COMMITTEE ON SCIENCE

F. JAMES SENSENBRENNER, Jr., Wisconsin, *Chairman*

SHERWOOD L. BOEHLERT, New York
HARRIS W. FAWELL, Illinois
CONSTANCE A. MORELLA, Maryland
CURT WELDON, Pennsylvania
DANA ROHRABACHER, California
JOE BARTON, Texas
KEN CALVERT, California
ROSCOE G. BARTLETT, Maryland
VERNON J. EHLERS, Michigan**
DAVE WELDON, Florida
MATT SALMON, Arizona
THOMAS M. DAVIS, Virginia
GIL GUTKNECHT, Minnesota
MARK FOLEY, Florida
THOMAS W. EWING, Illinois
CHARLES W. "CHIP" PICKERING, Mississippi
CHRIS CANNON, Utah
KEVIN BRADY, Texas
MERRILL COOK, Utah
PHIL ENGLISH, Pennsylvania
GEORGE R. NETHERCUTT, Jr., Washington
TOM A. COBURN, Oklahoma
PETE SESSIONS, Texas
VACANCY

GEORGE E. BROWN, Jr., California RMM*
RALPH M. HALL, Texas
BART GORDON, Tennessee
JAMES A. TRAFICANT, Jr., Ohio
TIM ROEMER, Indiana
JAMES A. BARCIA, Michigan
EDDIE BERNICE JOHNSON, Texas
ALCEE L. HASTINGS, Florida
LYNN N. RIVERS, Michigan
ZOE LOFGREN, California
MICHAEL F. DOYLE, Pennsylvania
SHEILA JACKSON LEE, Texas
BILL LUTHER, Minnesota
DEBBIE STABENOW, Michigan
BOB ETHERIDGE, North Carolina
NICK LAMPSON, Texas
DARLENE HOOLEY, Oregon
LOIS CAPPS, California
BARBARA LEE, California
BRAD SHERMAN, California
VACANCY

TODD R. SCHULTZ, *Chief of Staff*
BARRY C. BERINGER, *Chief Counsel*
PATRICIA S. SCHWARTZ, *Chief Clerk/Administrator*
VIVIAN A. TESSIERI, *Legislative Clerk*
ROBERT E. PALMER, *Democratic Staff Director*

SUBCOMMITTEE ON SPACE AND AERONAUTICS

DANA ROHRABACHER, California, *Chairman*

JOE BARTON, Texas
KEN CALVERT, California
ROSCOE G. BARTLETT, Maryland
DAVE WELDON, Florida
MATT SALMON, Arizona
THOMAS M. DAVIS, Virginia
MARK FOLEY, Florida
CHARLES W. "CHIP" PICKERING, Mississippi
CHRIS CANNON, Utah
KEVIN BRADY, Texas
MERRILL COOK, Utah
GEORGE R. NETHERCUTT, Jr., Washington

BART GORDON, Tennessee
RALPH M. HALL, Texas
JAMES A. TRAFICANT, Jr., Ohio
ALCEE L. HASTINGS, Florida
ZOE LOFGREN, California
SHEILA JACKSON LEE, Texas
BILL LUTHER, Minnesota
NICK LAMPSON, Texas
LOIS CAPPS, California
TIM ROEMER, Indiana

---

*Ranking Minority Member
**Vice Chairman

CONTENTS

NASA AT 40: WHAT KIND OF SPACE PROGRAM DOES AMERICA NEED FOR THE 21st CENTURY?

THURSDAY, OCTOBER 1, 1998

U.S. HOUSE OF REPRESENTATIVES,
COMMITTEE ON SCIENCE,
SUBCOMMITTEE ON SPACE AND AERONAUTICS,
Washington, DC.

The Subcommittee met, pursuant to notice, at 12:12 p.m., in room 2325, Rayburn House Office Building, Hon. Dana Rohrabacher, Chairman of the Subcommittee, presiding.

Chairman ROHRABACHER. All right. Ladies and gentleman, face to face with the future is what this hearing's about. I hereby call this meeting of the Space and Aeronautics Subcommittee to order.

Today is October 1, 1998, but we can also feel the spirit of October 1, 1958, and, perhaps, even that of October 1, 2038. I say this because today we are celebrating the past; we are assessing the present; and we are anticipating the future of America in space. Today, we will see through the eyes and the words of five leading authorities on where America can, should, and, perhaps, must go in space. And thanks to this brand new videoconferencing technology that's being demonstrated here today, we are actually looking the future in the face, and that will be part of our hearing. As we go on, we'll have Pete Conrad testifying from California through this technology.

Today, we are beginning a national dialogue about the future of America's space program, and we need to bring the American people into this discussion, not just because we're spending their money, but because we're talking about their future and their children's future. In the past, space policy has been a fairly top-down process. The Presidents were proclaiming new goals or cutting the budget, Congress working its will, and NASA managing and trying to implement what was being told to them. In the future, I envision space will be a frontier that is open to all—stockholders as well as scientists, educators as well as engineers, tourists as well as technicians. All will be able to participate in America's space enterprise. They should, therefore, participate fully in setting the course to reach that visionary future.

Space must not simply be a government program of a wealthy Nation, but, instead, it should be the engine of new knowledge, prosperity, and peace that will strengthen our democracy, expand our economy, and spread our democratic values, opening horizons of human opportunity beyond the confines of planet Earth. There are so many new ideas and miraculous possibilities, and it's exciting to be part of this great endeavor, and it's exciting for me to be the Chairman of this Subcommittee at this pivotal moment.

Through NASA, we got our start in space 40 years ago, but, as Ronald Reagan used to say, "You ain't seen nothin' yet." Or, as he said, "You ain't seen nothin' yet."

[Laughter.]

Chairman ROHRABACHER. So, with that, I'd like to turn to my Ranking Member, Mr. Gordon, for an opening statement.

Mr. GORDON. Thank you, Mr. Chairman. First, I would like to ask unanimous consent that Mr. Brown's opening statement be entered into the record as well as written testimony submitted by the Planetary Society and by Mrs. Marcia Smith.

Chairman ROHRABACHER. And there's certainly no objection, so ordered.

[The prepared statements of Mr. Brown, Ms. Smith, and The Planetary Society follow:]

OPENING STATEMENT

by

HON. GEORGE E. BROWN, JR.

October 1, 1998

Good morning. I would like welcome our witnesses to today's hearing. I look forward to your testimony. I also would like to congratulate Administrator Goldin on NASA's 40th anniversary. It is a significant milestone, and you and all of the present and former NASA and contractor employees should feel proud of what our nation's civil space program has accomplished. I believe that space exploration is a noble calling, and one that is worthy of our continued support.

When I celebrate a birthday or anniversary, I like to reflect both on what has come to pass in my life and what might still lie ahead. I think that it is appropriate that we do the same at this juncture in NASA's life as an institution. Thus I am pleased that the subcommittee is holding today's hearing.

It is difficult to know what to say and what to leave out in an opening statement such as this. I have been interested in our nation's civil space program since its inception, and I have rejoiced at its successes and grieved at its setbacks over the years. I also have tried to play a small part in helping to nurture our space program's growth during my tenure in Congress.

As we try to assess how best to approach America's future in space, I would like to offer some thoughts on ways in which our civil space program has succeeded, as well as ways in which it has fallen short. I hope that those observations may provide useful guideposts for NASA's future.

Where has our civil space program succeeded? I believe that there can be no disputing that the nation's history of excellence in space and Earth science is one of the most gratifying outcomes of the space program's establishment 40 years ago. To say that we have been unlocking the secrets of the universe may be a cliché—but it also is a simple statement of fact. Robotic probes have visited every other planet in our solar system except Pluto. Equally important, we are gaining insights into our own planet and into how human activities affect its systems.

Yet, as the news about the immense burst of gamma-rays and X-rays from deep in space that bathed Earth in radiation last month and disrupted spacecraft operations demonstrates, we also are learning how *little* we really know about the universe. Nevertheless, we *are* making advances. From the earliest satellites to the

Hubble Space Telescope, the Mars Pathfinder, and beyond, NASA's scientific activities have enriched our stores of knowledge and captured our imaginations. I find it incomprehensible that anyone could seriously asert that NASA has made these space activities "boring".

Another area where I think our space program has succeeded is in the use of space technology to deliver benefits to society. Communications satellites, weather and remote sensing satellites, technologies for health care and industrial competitiveness, and so forth—all of these have improved the quality of our lives. Yet I strongly believe that we can do more to harvest the potential of space to help people in America and around the world to live better lives. We dare not walk away from that responsibility.

Another development that has been gratifying to me has been the emergence—albeit somewhat unevenly—of a healthy and growing commercial space sector. Can anyone dispute the importance of commercial satellite communications to the world's economy, both directly and indirectly? I expect to see similar developments in commercial remote sensing, launch services, and in areas we do not even contemplate at present.

Yet, a world with a large, global commercial space sector offers challenges to policymakers. I would gently note that those who argue for unleashing the private sector to commercialize space on a global scale should not then attempt to demonize companies for pursuing their interests globally. Loose allegations of unpatriotic behavior against American companies are not helpful as we attempt to sort out the complex issues associated with commerce in a multipolar world.

Next, I must say that I consider one of NASA's finest achievements to have been its continued adherence to the principle that *"activities in space should be devoted to peaceful purposes for the benefit of all mankind"*. NASA has made international cooperation an integral part of its mission, and I think it has paid dividends to our nation over the past four decades. As I testified to this subcommittee fifteen years ago on NASA's 25th anniversary:

"We have tended to cut back on these activities when budgets are tight, or when we get nervous about technology transfer to other nations. My view is that international cooperation serves many functions and is important to support; political, scientific, and economic benefits can be derived from international cooperation."

I continue to stand by those sentiments.

Before I close, I must confess that I have been disappointed by what we have *not* accomplished in our space program since its inception. Among my frustrations is our sorry record in the development of low-cost space transportation. For too long, we have seen advances in transportation come in fits and starts, instead of being an integral part of the space agency's R&D portfolio. That approach has seriously impeded the progress of our nation in space exploration and utilization, I believe that we are making some improvements in our approach to developing new transportation systems; however, we must not go overboard in the other direction and make our programs dependent on the delivery of revolutionary advances by a date certain. Innovation does not work that way—especially under constrained budgets.

Speaking of budgets, those who know me know that I long have been troubled by the nation's inconsistent support of our civil space program. Both Congress and the White House have at various times over the last forty years, wreaked havoc on NASA's ability to make and carry out long-range plans. While some have argued that there is no long-term vision or plan for NASA, I would submit that over the years there have been no lack of plans, studies, and recommendations both from the space agency and from distinguished panels of outside experts—my files are filled with them! Rather, the nation—and I most definitely include Congress in this criticism—has been unable or unwilling to establish institutional mechanisms for providing a stable budgetary and programmatic framework in which long-term (i.e., over multiple Administrations and Congresses)—and possibly multinational—R&D initiatives can be carried out successfully. Until we address that situation, I am not optimistic that we will be able to achieve many of the ambitious goals that NASA is contemplating for the 21st century.

One of those goals, of course, is human exploration beyond low Earth orbit. I believe that this is one of the areas in which we have fallen most short, to my intense disappointment. I still consider it extremely short-sighted for the Nixon Administration to have canceled the nation's program of lunar exploration just as it was transitioning from an amazing technological and managerial achievement to a sustained program of scientific and human exploration.

Then, when President Reagan approved the development of a Space Station in 1984, I never would have believed that almost fifteen years later we still would not have launched even the first elements of the Station into orbit. Still, there are no easy villains in the Space Station saga—as much as some would like to find them. The simple truth is that NASA, successive Congresses, and successive Administrations all have had their fingerprints on this program, and any blame for the slow progress of this program needs to be shared. At the same time, all of us will have to now work together for this program to be completed successfully.

However, I would like to look forward to the future of human space exploration and not just dwell on past missteps. As Lou Friedman, Executive Director of the Planetary Society and Marcia Smith, former Executive Director of the National Commission on Space point out in the thoughtful testimony that they have submitted for this hearing (and for which I would ask unanimous consent that

it be entered into the record), there are no lack of worthy goals for human exploration. Those goals need to be a fundamental part of our vision for NASA's future.

While there is more that I could say, I will conclude my statement on that note. I have offered these observations with the intention of stimulating some discussion on how best to proceed as our space program enters the 21st century. We must learn from our past if we are to avoid becoming captive to our past.

Thank you.

NASA AT 40

HEARING BEFORE THE HOUSE SCIENCE COMMITTEE SUBCOMMITTEE ON SPACE AND AERONAUTICS

STATEMENT FOR THE RECORD

MARCIA S. SMITH
FORMER EXECUTIVE DIRECTOR
NATIONAL COMMISSION ON
SPACE

OCTOBER 1, 1998

NASA at 40

Statement for the Record

Marcia S. Smith[1]
Former Executive Director
National Commission on Space

Mr. Chairman, Members of the subcommittee, thank you for offering me the opportunity to present my views regarding the future of the civilian space program on the occasion of NASA's 40th birthday.

This testimony is presented from my perspective as former executive director of the National Commission on Space (NCOS). NCOS was created by Congress in 1984 as part of the FY1985 NASA authorization bill. In that same bill, Congress approved initiation of the space station program. NASA had sold the space station program as the "next logical step" in the space program, yet when Congress asked "the next logical step to what?", the answer was not readily apparent. NCOS was created to develop that answer.

Congress directed President Reagan to name our 15 commissioners and gave us one year to complete our task. We opened our doors in June 1985 and closed them in July 1986. Tom Paine, NASA Administrator during the first Apollo missions to the Moon, was selected to chair the commission. Dr. Paine, who passed away several years ago, was a fervent space visionary and our report reflects his enthusiasm, vivacity, and determination. Our commissioners came from a wide variety of backgrounds, and several in addition to Dr. Paine fit the sobriquet of "visionary." (The names of all the commissioners are listed at the end of this statement.) We also had help from many other professionals in the space community, as well as input from taxpayers around the country through a series of 15 town meetings, letters we received in response to a Parade magazine article about our efforts, and e-mails over the nascent computer network Compuserve.

In the legislation that created the Commission, Congress asked us to look out 20 years into the future. We decided to extend that time frame to 50 years, however, because in the space business, 20 years isn't a very long time. For example,

---

[1] Ms. Smith is currently a Specialist in Aerospace and Telecommunications Policy for the Congressional Research Service. The views expressed here do not represent those of CRS or of the Library of Congress.

Congress told us to take the space station and the space shuttle as "givens" and develop a space program for which those were the stepping-stones. NASA currently anticipates that the space station will be completed in 2005, exactly 20 years from when we started our deliberations.

So it's good that we expanded our outlook to the year 2035. The space station was supposed to have been completed in 1994. Our report would have taken the country 40 years past that time. It still may serve as a template for doing that.

OUR RECOMMENDATIONS

For those who remember the Paine Commission, they undoubtedly recall our human spaceflight recommendations because they garnered the most attention. We recommended that humans return to the Moon by 2005 and land on Mars by 2015. But we also made many other recommendations that span the full range of civilian space activities. We proposed three mutually supporting thrusts:

- Advancing our understanding of the planet, our Solar System, and the Universe;
- Exploring, prospecting, and settling the Solar System; and
- Stimulating space enterprises for the direct benefit of the people on Earth.

To accomplish those recommendations, we said that America had to make a long-range commitment to two additional thrusts:

- Advancing technology across a broad spectrum to assure timely availability of critical capabilities; and
- Creating and operating systems and institutions to provide low-cost access to the space frontier.

Many considered those to be bold objectives, but, in fact, we were much bolder than that!

> OUR OVERARCHING GOAL WAS THAT THE UNITED STATES SHOULD LEAD THE WAY IN OPENING THE INNER SOLAR SYSTEM FOR SCIENCE, EXPLORATION, AND DEVELOPMENT.

Our entire approach to the future of space was bold even though the times in which we met advised caution. The U.S. government was in the throes of the Gramm-Rudman-Hollings budget balancing exercise, which was intended to reduce the budget deficit to zero within 5 years. Though many were skeptical it could be achieved, we decided to take it at face value. At that time, NASA's budget was approximately one-quarter of one percent of the gross domestic product (GDP). We assumed that by the 1990s, the U.S. economy would be sufficiently strong and the budget sufficiently under control that NASA's budget could gradually increase to one-half of one percent of the GDP by the turn of the century. At that higher level, the nation would be spending half what it spent on civilian space during the Apollo years as a percentage of GDP.

We also assumed that commercial space would thrive and companies would be investing in space. With that combined level of investment, plus the involvement of other countries, we laid out a robust series of commercial activities and robotic scientific programs, and created a network of transportation systems and space stations to allow people routine access to the Moon and Mars — a "highway-to-space" and a "bridge-between-worlds."

Our human spaceflight recommendations unquestionably were infrastructure-intensive and I'm not certain any of us thought that such a network ever would be built. The point was to design a system that would avoid another dead-end project like Apollo. We specifically rejected the concept of "sprint" missions to Mars where a few astronauts would pick up a few rocks and come home as we did during Apollo. As expensive and time consuming as our approach was, it would ensure that the full potential of the Moon and Mars could be realized for scientific research, resource utilization, and exploration. Achieving that goal was the point, not following a particular blueprint developed in the mid-1980s. Hopefully, technological advances will show us easier, more cost-effective ways to reach that goal.

We also were bold in the face of one of the great tragedies of our space program. Midway through our term, the space shuttle Challenger exploded. We were as deeply affected as anyone by that terrible loss of life and we did stop to reconsider our conclusions. Ultimately, though, we decided that the space shuttle would resume service someday, and that while the Challenger accident might affect our timetable, it did not change the overarching goal. We believed that the American people would not shrink away from that setback, but would rally behind NASA and resume sending people into space once appropriate fixes were made. We were right.

WHY AREN'T WE THERE YET?

Twelve years later, the bold future we expounded is making mixed progress. In the scientific arena, NASA is making important strides in studying the Earth, exploring the solar system with robotic spacecraft, and advancing our understanding of the Universe. Commercial use of Earth orbit, primarily for communications and remote sensing, has expanded considerably since our report was published, while other commercial activities, such as materials processing in space, are developing more slowly than hoped. The areas that have shown the least progress are human spaceflight and developing low cost access to space.

The pace of progress in human spaceflight has been slowed not by the Challenger tragedy, but because the budget situation took several years longer to resolve than anticipated, and none of us imagined that the space station would take so long to build. Congress had told us to take NASA's space station as a given and we did, assuming it would be completed in 1994 as promised. We recommended building additional space stations in various orbits, particularly a "variable gravity" space station in Earth orbit that would rotate around a tether to create different levels of gravity (g). Such a facility would enable studies of how humans adapt to various space environments like the Moon's 1/6 g and Mars' 1/3 g, in addition to the near-zero g in Earth orbit. But NASA's space station was the foundation we were told to build upon, and it is still missing.

In the 12 years since our report was published, the completion date for the space station has slipped repeatedly. The difficulties faced in that program make many cautious about initiating new "big" space programs. The United States had a space station in 1973-1974. The Russians have had 7 successful space stations since 1971. How can it be so difficult, costly, and time consuming to build a space station?

The most important step that needs to be taken to move the United States forward towards achieving the Commission's goal is to develop a comprehensive understanding of what has gone wrong with the space station program.

There are many factors to consider. Congress and NASA have struggled over keeping the space station's costs within various promised NASA funding envelopes. As costs grew, NASA would redesign the station to lower those costs. Redesign followed redesign. Cost estimate followed cost estimate. Fourteen years after the program began, NASA is still attempting to ascertain just how much the program will cost and how long it will take to build. The process of estimating costs, reaching agreement with Congress and the White House on those costs, and managing the program to remain within the promised costs, are all steps that will be crucial to the success (or failure) of the next big space project.

The advantages and disadvantages of international cooperation also need to be reexamined. Russia bears some—but not all—of the responsibility for the situation in which the space station program finds itself today. Russia over-promised what it could afford and the United States accepted those promises despite mounting evidence that Russia could not perform. But Russia is not the only space station partner that has encountered political or funding difficulties. Support for the program in the United States itself has fluctuated over the years. The European Space Agency and Canada have had to scale back their contributions. It may be time to reevaluate international cooperation in large space projects that span decades. Does international cooperation really achieve cost savings? Should such projects have fewer participants to simplify management? How can political and funding support be maintained in each partner country throughout the lifespan of a project? What problems result when space projects are tied to foreign policy agendas?

Before the United States moves forward in meeting bold new goals, NASA must fully comprehend what went wrong with the space station program. NASA needs to learn from those mistakes before undertaking another big space infrastructure project.

A penetrating analysis by an outside group with expertise in management of government programs, perhaps the National Academy of Public Administration (NAPA), would be one avenue for beginning the "lessons learned" process. Such an analysis should begin immediately.

The space shuttle, the other stepping-stone we were given, returned to flight in 1988, 32 months after the Challenger tragedy. But the objective of significantly lowering the cost of getting into space is as elusive today as it was in 1986 when the Commission made its recommendations, or in 1972 when the shuttle program was approved specifically for that purpose. The abundant number of "space transportation architecture" studies already completed by NASA and the Department of Defense (DOD) have not brought us to the 10-fold reduction in launch costs that long has been the stated goal of those who propose new launch systems. One would think enough studies have been done already, yet more are beginning at NASA today.

With a number of private sector companies proposing new launch vehicles, NASA's Reusable Launch Vehicle program, and DOD's Evolved Expendable Launch Vehicle program, progress hopefully will be made soon on lowering launch costs, though probably not by a factor of 10. To reach that objective, should the government invest more money in space launch technology because it would ultimately lower costs for government activities in space? Or should the companies building launch vehicles be expected to invest their own funds instead of using taxpayer money, since the commercial market for launch services is thriving? What are the strengths and weaknesses of government/industry partnerships?

As with the space station program, comprehending the technical and policy hurdles that have prevented the United States from developing low cost access to space is more important than another space transportation architecture study. Technical issues are important, but policy disagreements over the respective role of the government versus the private sector and the roles of the various government agencies involved also may have had significant influence on the lack of progress in achieving this goal.

NAPA or the National Research Council (or both) could be tasked to dissect the problems in achieving low cost access to space and identify solutions to this decades-old issue.

IS IT TIME FOR ANOTHER NCOS?

The National Commission on Space shaped a program to take us 40 years past the time when the space station started operations. Those goals and objectives haven't changed. Just the timetable.

Ours was the first of several studies over the past decade and a half about the future of the government civilian space program. One year after our report was published, NASA did its own study under the leadership of Sally Ride. That was followed by the "Augustine Report," chaired by Norman Augustine, then CEO of Martin Marietta, and several more by the National Space Council throughout President Bush's administration. It seems that a sufficient number of studies have been done to establish guideposts for NASA's next decades in space.

As the new century dawns, attention may shift from governmental space activities to the expanding commercial presence in space. Over the next decade, companies are likely to replace governments as the dominant users of Earth orbit and may initiate their own exploration beyond. Developing a better understanding of the most productive relationship between the U.S. government and the private sector in exploring and using space would be beneficial. What activities must governments perform because only they have the requisite resources? What activities can governments stimulate, with the expectation that the private sector will take them to completion? Where should the private sector invest its own resources without expecting help from the government? What government regulations are essential, and which create obstacles to commercial success?

As NASA celebrates its 40th birthday, thoughts turn to what NASA has accomplished in its first 40 years and where it should focus for the next four decades. Some ask whether it is time to gather another group akin to the National Commission on Space.

With the large number of studies on the future of civilian space activities that already have been completed, it is not clear that another National Commission on Space is needed yet. If one were to be established, however, it could address government-private sector relationships in an era of rapidly expanding commercial use of space.

KEEPING AN EYE ON THE FUTURE

Meanwhile, the key is to keep our eyes on the future and not be discouraged with present-day setbacks. We need to learn from past mistakes and then move boldly forward. It matters less when we start, or whether we follow a dusty 1986 blueprint, than that we try. As President Bush stated on the 20th anniversary of the first human landing on the Moon:

> "WHY THE MOON? WHY MARS? BECAUSE IT IS HUMANITY'S DESTINY TO STRIVE, TO SEEK, TO FIND. AND BECAUSE IT IS AMERICA'S DESTINY TO LEAD."

MEMBERS OF THE NATIONAL COMMISSION ON SPACE

(Positions shown are those held at the time the individual served on the Commission unless otherwise noted. Asterisks indicate those who are now deceased.)

*Dr. Thomas Paine, Chairman. Former NASA Administrator; former President of Northrop Corp.

Dr. Laurel Wilkening, Vice Chairman. Vice President of Research and Dean of the Graduate School, University of Arizona.

*Dr. Luis Alvarez, Nobel Laureate Physicist, Lawrence Berkeley Laboratory.

Mr. Neil Armstrong, Chairman, CTA Inc. First man to walk on the Moon.

Dr. Paul Coleman, President, Universities Space Research Association; Assistant Director, Los Alamos National Laboratory; Professor, UCLA.

Dr. George Field, Senior Physicist, Smithsonian Astrophysical Observatory; Professor, Harvard University.

Lt. Gen. William Fitch (Ret.), Defense Consultant; former head of Marine Corps Aviation.

Dr. Charles Herzfeld, Vice Chairman, Peterson, Jacobs, Ramo & Co; former Vice President, ITT Corp.; former Director, Advanced Research Projects Agency.

Dr. Jack Kerrebrock, Associate Dean of Engineering, MIT; former Associate Administrator, NASA.

Amb. Jeanne Kirkpatrick, Professor, Georgetown University; Senior Fellow, American Enterprise Institute; former U.S. Representative to the United Nations.

*Dr. Gerard O'Neill, President, Space Studies Institute; CEO, Geostar Corp.

Gen. Bernard Schriever (Ret.), Management Consultant, Schriever & McKee Inc.; former Commander, Air Force Systems Command; former Director, Air Force Ballistic Missile Program.

Dr. Kathryn Sullivan, NASA astronaut; Adjunct Professor, Rice University. First American woman to walk in space.

Dr. David Webb, Consultant in the field of space development; Chairman, National Coordinating Committee for Space.

Brig. Gen. Charles (Chuck) Yeager (Ret.), Test Pilot. First man to break the sound barrier.

NASA: Exploring New Worlds

Testimony for the House Science Subcommittee on Space and Aeronautics

By Dr. Louis Friedman
Executive Director, The Planetary Society

October 1, 1998

Mr. Chairman, Ranking Member Gordon, Members of the Subcommittee, I am pleased to present this statement on behalf of The Planetary Society – the largest space interest organization in the world, with100,000 members in over 140 countries. It is appropriate that the Congress and the public consider, on this 40th anniversary of NASA, America's vision of and aspirations for achievement in space exploration.

NASA stands as a beacon of hope and symbol for a positive future for not just America, but the whole world. People everywhere are justifiably proud of the achievements of the past 40 years, achievements which continue into the present and bode well for the future. Whereas the agency was born in response to a perceived enemy challenge in a cold war, the legacy of the ensuing space race was not enmity between superpowers, but a sense that "we came in peace for all mankind." Gradually the emphasis shifted from nationalism to internationalism and cooperation, as appropriate for a planet exploring new worlds.

All of us today should appreciate the foresight of those who formed NASA and provided for a civilian agency whose main mission was to explore and search for knowledge. NASA has come to symbolize the value of excellence, the importance of achievement, the significance of scientific knowledge, and the wonder of exploration for the entire world. In the first two decades of its existence, not only did they put humans on the Moon, but they visited every planet known to the ancients, landed on Mars, and sent four spacecraft on journeys toward other stars. Unfortunately, they bogged down in their third decade – turning their sights inward, remaining hidebound on Earth or in Earth orbit, pursuing illusory notions of routine access to space, commercial benefits, and near-term return on investment – and forgetting the spirit of adventure and search for knowledge that were the *raison d'etre* for America's national space program.

But in the fourth decade pride in the agency has returned -- the **Hubble Space Telescope** looks to the edge of our universe; **Pathfinder** took us on a new journey to Mars and captured the hearts and spirit of humanity in a way we never thought a little robot could; **Mars Global Surveyor** continues to reveal an alien world that holds many clues as to the nature of life; NASA scientists analyzed a piece of Mars that fell on Earth and found clues suggesting past life on that planet; there have been hints of a warm underground ocean on Europa; and insights into the swarm of objects in Earth's neighborhood that have shaped the evolution of our planet and the life on it. America is exploring again – and pride in NASA has been restored, worldwide.

NASA is now attempting the grandest and largest international engineering project in history – the international space station. Proposed first in the dissolute 80s, it went nowhere until it was made truly international and American astronauts flew missions on the **Mir.** That shuttle-**Mir** program was conducted brilliantly, and added immensely to American capabilities in space. The development of the space station is now in trouble – pushed and pulled by political and economic forces greater than those of physics and engineering. That great ventures will have great difficulty is no surprise – the challenge is now to decide whether it remains a great venture worth the cost, or whether it becomes diminished to a point that its very rationale is lost. It can be the former only if it serves international political and social purposes commensurate with the cost, and if it is part of an exploratory venture that leads somewhere in the future. As an end in and of itself, it will not be worth it, and it will bog down the space program in a repeat of that sad third decade of NASA's in the 1980s.

The Planetary Society represents a public viewpoint – our members are not the professionals who build the spacecraft and rockets or who analyze the data. They are the citizens of the world who pay for and want their leaders to spend tax money on space missions. The motivation is not practical benefits: people don't care if their communications is via satellite or fibre optics; or where their weather data come from – the practical and commercial will be solved by the private sector and the marketplace. We invest government money in space for a different reason – to seek knowledge, to explore new worlds, to search for other life, and to understand ourselves, our origins and destiny. That spirit of exploration and discovery must remain the focus of the space agency and the government space program.

That was forgotten a few years ago. The United States had its astronauts in orbit, no destination beckoning them. America had been to the Moon, but we went no further. NASA found that another planet may have supported life, or might in the future support life, but was afraid to venture to it. It took several blue-ribbon commissions (led by visionaries like Sally Ride, Tom Paine and Norm Augustine), and a new Administrator, to change that. A national space policy now commits us to sustained robotic presence on Mars and sample returns from the bodies of the solar system. The international space station has combined all the world's human spaceflight programs in a single effort to prepare humans to conduct long-duration space ventures – and the only space venture requiring long-duration is interplanetary flight. Mars, as the media continually reminds us, is the human destination people throughout the world all wonder about. Will humanity be a multi-planet species, or not? We must keep our eyes fixed on that goal. It may be on a distant horizon – but as Maya Angelou points out, "The horizon leans forward/ Offering you space to place new steps of change..." Not some distant horizon, to be sought or reached after "we've solved our problems here at home," but, she said, on a horizon leaning forward, beckoning us to do our best, "to place new steps of change/ Here on the pulse of this fine day."

There are many technical and scientific questions to be solved, and challenges to be overcome. Today's testimony should not be about the mundane and the intricacies of overseeing a government agency – you deal with that in the budget and oversight hearings. Whether human missions to Mars should take place in 2010, 2015, 2020; whether there should 2 or 3 robotic sample returns first; whether we have a human return to the Moon or set up test bases in the Antarctic, whether new technology make missions less costly – these are questions that NASA must deal with – when we, the public, have the will to ask them to do so. Today's hearing and testimony should reflect on the past 40 years, and set a vision for the next 40. As former Planetary Society Advisor, Norman Cousins said, "We can always do better than we are doing if we emancipate ourselves from our fears in order to search the horizon for new prospects.... The trip to Mars has to do with our loneliness in the Universe. We are transported by the notion that there may be other humans out there too.... Infinity converts that which is possible to that which is inevitable.... To have a rendezvous with infinity will be the ultimate human achievement."

Mr. GORDON. Good morning to everyone. As has been noted, today is the 40th anniversary of America's civil space program. For this important milestone, I think it's appropriate that we pause to salute the accomplishments of the National Aeronautics and Space Administration over the past 4 decades. And, of course, when we salute NASA, we really are honoring the dedicated scientists, engineers, administrators, and support staff in both the civilian service and the contractor community who have achieved so much.

I imagine that it would have been hard to believe in 1958, just a year after the Soviet Union successfully orbited Sputnik while the U.S. space rockets were blowing up on the launch pad, that the U.S. space program would have come so far. When we look back from the perspective of 40 years later, NASA has had many successes that have advanced our knowledge of the Earth and the rest of the universe and have helped humanity take its first steps onto another world and have provided a whole series of practical benefits to our citizens back here on Earth.

It has not been an unblemished record, and this Committee has, over the years, held NASA accountable for its failures when appropriate. We owe the hard-working American taxpayers no less than that. Yet, today should not be a day of dwelling on NASA's shortcomings. There will be another opportunity for such oversight when needed.

Instead, today should be a day for recognizing the civilian space program's achievements and reflecting on what America wants to accomplish in space over the next 40 years. In that regard, I welcome our witnesses to today's hearing and look forward to their testimony. Thank you.

Chairman ROHRABACHER. All right, without objection, the written statements of other members will be made part of the written record. Hearing none, so ordered.

At this point, we will go through the formality of swearing in the witnesses. We have five fine witnesses, and one who, through technology, will be joining us as well, but for those who are here, could you please stand? I will administer the oath.

If you would raise your right hand. Do you swear that the testimony you are about to give is the truth, the whole truth, and nothing but the truth?

Mr. GOLDIN. I do.

Mr. CONRAD. I do.

Mr. MCCURDY. I do.

Ms. GALLOWAY. I do.

Mr. TUMLINSON. I do.

Chairman ROHRABACHER. Okay. The reporter shall note that each witness responded in the affirmative. You may be seated.

Since we have a full panel today, I'd like to ask each of you to summarize your testimony, and there will be probably several rounds of questions, and we will be getting into some of the areas—and actually some areas that you might not touch on in your remarks. We're going to be asking you about or asking you to expand on certain issues, and we've got plenty of time for questions, but if you could summarize what you have down into 5 minutes, it would be very helpful to the Committee.

And, first up, we have with us someone who is, of course, very familiar to everyone in the room and the Subcommittee, Dan Goldin, Administrator of NASA, and, you know, a lot of times when Dan is up here people forget he's got all this authority and power and all this influence here now, but let us also remember that this was a man who was very successful in the private sector before he came to Washington, DC. to loan us his talents and his skill and his time which, as we know, is—anyone over 50 knows that loaning us your time is really a valuable thing.

[Laughter.]

Chairman ROHRABACHER. We are very, very lucky to have Dan Goldin willing to give us his time and his energy and his skills and his talents to help make America's space program what it should be, and sometimes I give Dan a hard time, even me, but even then we all—I think Dan has a greater degree of respect than any person that I've seen heading NASA in my lifetime, and it's really great to have you with us, and I'm very proud to call Dan Goldin my friend and my surfing partner.

[Laughter.]

Mr. GOLDIN. You're a better surfer than me.

Chairman ROHRABACHER. I won't tell anybody that, it's really not a big accomplishment, but anyway, Dan——

[Laughter.]

Chairman ROHRABACHER. But, with that, Dan, you can proceed.

TESTIMONY OF DANIEL S. GOLDIN, ADMINISTRATOR, NATIONAL AERONAUTICS AND SPACE ADMINISTRATION

Mr. GOLDIN. Mr. Chairman, members of the Committee, it is with tremendous pride that I am here this afternoon to testify on the occasion of NASA's 40th anniversary. NASA has had a great 40 years, but we are just getting started. We believe that there will forever be something to invent, somewhere to discover, and someplace to visit. The work to find out what and where, that has already begun. My written statement goes into detail into what we hope the future holds for NASA and our country, but in my limited time this afternoon, I'd like to summarize the NASA vision.

It is my hope and dream that we could erase any and all distinction between aircraft and spacecraft. It is my hope and dream that thanks to revolutionary breakthroughs in advanced materials and coatings, distributed intelligence and control systems, and advanced lightweight structural concepts integrated with the next generation of air-breathing jets and rockets that we could fly a vehicle routinely from Washington, DC. to Japan in a little over 2 hours. It is my hope and dream that high definition synthetic vision that allows the pilot of this vehicle to clearly see through even the heaviest fog, snow, or dust and advanced crew interface that gives our pilot real-time, on-board, unprecedented air traffic control to make this flight much safer than ever before.

It is my hope and dream that we have advanced remote sensing technologies so far that we could make the most accurate predictions about climate and weather and natural resources. It is my hope and dream that these are not 5-day predictions or even monthly predictions, but, rather, seasonal to annual to multidecadal predictions that truly promote sustainable development for future generations.

It is my hope and dream that we revolutionize telescopes so that they have significantly better spatial and spectral resolution than their predecessors at a significant fraction of the weight and an even larger fraction of the cost. We intend to directly detect Earth-size planets around stars within 100 light years of Earth; that's 600 trillion miles. If these planets exist and we're able to isolate them, we'd like to take a picture with a resolution high enough to see biological signs and geological features. The last 40 years—and, by the way, think of a picture right behind you that's not of Earth looking at the moon but is of another planet around another star looking the other way. Think about what that would do for our children.

The last 40 years have been about taking a census of our solar system. It is my hope and dream that the next 40 are about taking a census of interstellar space. It is my hope and dream that these important missions will help us answer the age-old questions: What are the origins of the universe? How did galaxies and stars and planets evolve? Does life in any form exist elsewhere than on planet Earth? Are we alone? It is my hope and dream that we'll be able to answer those profound questions because we realized and took advantage of the biological revolution that took place in the latter half of the 20th century.

It is my hope and dream that we understand that if we were committed to looking for life on other planets, we needed a fundamental understanding, an intellectual underpinning of biology, so we knew what kind of life to look for. But just as important, it is my hope and dream that we realize, because the questions we sought to answer were so fundamental and far-reaching and because the challenges were so great, we took this intellectual underpinning of biology and applied it not just to our science but to our computing and engineering. The brain is a million times faster than our most powerful computer, and a single cell of the body could store 10 times more information than the most powerful computer. Just think about the potential if our computers can match that capability.

It is my hope and dream that this didn't only help us meet our goals in aeronautics, Earth science, and space science, it is my hope and dream that not only did American industry thrive with these new technologies in hand, Mr. Chairman, it is my hope and dream that these technologies allowed NASA to leave low-Earth orbit and take the next step in exploration and discovery. It is my hope and dream that commercial entities using their own state-of-the-art, low-cost, reliable space vehicles are mining asteroids and comets for valuable resources.

It is my hope and dream that there will be a colony on the moon with an industry-built infrastructure packed with hotels and a thriving tourism industry. It is my hope and dream that the International Space Station, in addition to leading to all kinds of biomed and industrial discoveries, prepares us so we could build a transportation node perhaps at the solar L–1 point, the balance point where the force of the sun and the Earth is equal. It is my hope and dream that node would be a base for our telescopes, both those looking at Earth and outer space, and a starting point for robotic voyages into intersolar and interstellar space, both orbiters and landers, and a launch site for our human missions, both to the planets and the moons of our own solar system.

It is my hope and dream that an astronaut in a white spacesuit with an American flag on their shoulder steps out of the spacecraft and crunches their boot in the dusty red surface of Mars and plants the American flag.

Mr. Chairman, before I conclude my remarks, allow me to come back to the present for a moment to focus on one of the most important requirements that will enable our visions: space transportation. After 40 years, it should be quite clear that we should never place our faith in a single launch system. The broad array of research and exploration goals that we want to pursue will go nowhere unless we have significantly more reliable and affordable transportation capabilities not controlled by the government. It has taken us many years, but we are now in a position where we are able to rely on one launch system while we make progress on an advanced capability. At times, this dual approach seems to generate unnecessary competition among our own team members.

Both the Space Shuttle and the Reusable Launch Vehicle (RLV) are vital to the future of America's space program. The Shuttle has been an important part of our history and until a proven replacement is able to be rolled to the launch pad, this is the vehicle that will guarantee human access to low-Earth orbit and play a critical role in the assembly of the international space system. The Shuttle system requires the dedicated attention of hundreds of civil servants and thousands of support contractors for operations and maintenance. The members of the Space Shuttle team have refined and improved this vital transportation system throughout its operational life. The Shuttle team has compiled an enviable record of research and operational accomplishments. Additionally, they remain the benchmark for excellence for air and space safety.

After the International Space Station is assembled, it is our intention to transfer the research and transportation requirements from the Shuttle to the RLV. It is my hope and dream that the launch vehicles developed both with and without government support will provide a much higher degree of reliability and operability that will provide dramatic reductions in the cost of access to space. With less money directed toward Shuttle operations, NASA will be able to direct more of its scarce resources towards long-range research and development activities.

It is my hope and dream that the Shuttle team members will be retained for new cutting edge assignments, and, no doubt, many will seek responsibilities with the growing Reusable Launch Vehicle industry.

In closing, let me say that it is my hope and dream that we do nothing less than open the air and space frontiers, and it is my hope and dream that our children will follow our work, and we will be inspired to take even bolder missions than our parents did. It is my hope and dream that exploring space enriches life here on Earth in ways we cannot even begin to predict. That's where NASA wants to go. That's what NASA wants to do.

Mr. Chairman, just as Congress and every Administration has supported NASA in the past, it is my hope and dream that you support the NASA of the future, because if you do, with the dedicated and determined people who work at NASA, these won't be hopes and dreams for long. Thank you very much.

[The prepared statement of Mr. Goldin follows:]

Statement of

Daniel S. Goldin
Administrator
National Aeronautics and Space Administration
before the

Subcommittee on Space
Committee on Science
House of Representatives
October 1, 1998

Forty years ago, in 1958, the National Aeronautics and Space Administration was created with the boldest and most noble of missions: to pioneer the future. We were told to explore new frontiers and enhance life here on Earth. We were asked to instruct; we were expected to inspire. Forty years later, thanks to an American public with an unquenchable thirst for knowledge and a relentless sense of adventure, NASA has delivered.

Think about this: forty years ago, jet passenger service was a novelty. Global communications meant a telephone line laid across the bottom of the Atlantic Ocean. When NASA was first getting started, the only way to track hurricanes was to fly planes directly over and into the storms. Our universe -- even the cosmic neighborhood just above our atmosphere -- was a mystery. In 1958, sending humans to the moon was pure science fiction.

But we dared to dream. We imagined what could be possible. And then along with our partners in industry and academia, we went to work.

In 1998, hundreds of millions of people ride American jets each year and new designs for flight go higher, faster and farther than ever before. Global space communications have helped create a global community. Weather satellites can track phenomena, like an El Niño condition, months in advance. There are still many mysteries to be solved, but Voyager, Galileo, the Hubble Space Telescope and other planetary and astronomy missions have circled neighboring planets, given us our first view of black holes, and begun to peer back at the very beginning of our universe. A space program that is forty years old has sent humans to the moon, robots to Mars, spacecraft to the furthest reaches of our solar system, and soon will help lead construction of the International Space Station. And for every step we take out there, we have contributed to a better quality of life right here. That is true whether it be one of thousands of "spin-off" technologies that has produced new products and services such as medical devices to detect breast cancer earlier. That is true when a child who looks up and knows that no longer is the sky the limit; it is the stars and beyond.

NASA has had a great forty years, but what the American people can be most proud of is this: when it comes to pioneering the future, we are just getting started. What will always define this aeronautics and space program -- and this country -- is our firm belief that there will forever be something to be invented, somewhere to be discovered, someplace to be visited. Rest assured, NASA will do its best in the next forty years to find out just what and where that will be.

In your invitation to be here today, you asked me to address NASA's strategic goals for the next 40 years. I welcome the opportunity to extend our reach beyond our near term challenges and ponder a future we may chose to create.

Keep in mind, I will be describing goals based on a vision that I believe is achievable in the long-term. This vision is not necessarily built on an mission profile that has been proposed through the traditional five-year budgetary process. Allow me to fuel your imaginations by describing a vision that a Congressional Committee may well hear about four decades from now. If we are able to align four critical factors including scientific and technical opportunities, a political imperative, national economic impact, and national spirit, I believe this vision can become reality.

It starts 25 years from now with a commercial flight from Tokyo to a commercial spaceport near the Kennedy Space Center. Thanks to revolutionary breakthroughs in advanced materials and coatings, distributed intelligence and control systems, and advanced lightweight structural concepts integrated with the next generation of air-breathing jets and rockets, the flight was completed in a little over two hours. Not bad!

Because of a thick fog, there is practically zero visibility at the spaceport. But that is not the concern that it was back in 1998. Now, high definition synthetic vision allows the pilot to clearly see through even the heaviest of fog, or snow, or dust. It allows the pilot to see better at night, as well. The vehicle also has an advanced crew interface. We have been able to couple humans and computers using natural, high order language to maximize performance.

The pilot doesn't control the vehicle with buttons, keyboards, and a joystick. Instead, the computer interface is an intelligent interchange with total immersion in sight, sound and touch. The computer is aided by waves from the pilot's brain. It monitors brain waves to determine -- and even alert the pilot -- when their attention might be lapsing.

For the entire two hour trip from Japan to Florida, the pilot has full situational awareness of the vehicle's state of health and the surroundings. Remote sensing satellites send signals directly to the vehicle and gives the pilot real-time knowledge about weather and terrain. GPS navigation systems, also readily accessible by the pilot, give the precise position. Communication links and onboard sensors determine where other vehicles and obstacles are, and an onboard computer database tells the pilot and crew where the other vehicles, obstacles and micro-disturbances in the atmosphere will be.

Together, this gives the pilot real-time, on-board, unprecedented air-traffic control. And all of this has made the flight safer and faster than ever before. This flight used to take 17 hours, not two, and the heavy fog would have had the plane circling or looking for another airport to land. But not today. And that's important.

It is important because the Japanese Prime Minister is joining members of the Administration and Congress, along with other heads of state, to welcome the return of an international crew returning from space. Deep space.

Their mission has been a great success. The crew just launched an interstellar probe from where they were stationed for the past few years: An international permanent outpost near Mars. The installation conducts research and operational activities supported by government and private sector investment.

By this time advanced telescopes have imaged countless Earth-like planets beyond our solar system with a resolution high enough to see geological and biological processes at 600 trillion miles. We've learned so much from watching our own planet as well as studying the other planets in our solar system, especially Mars and Europa, that we can now identity the fingerprints of life. We know there's life on Earth. We will find out if it is ubiquitous to the Universe.

So the probe our astronauts launched is going to travel to, remotely sense, and drop a mini robot on one of the planets we've discovered around a near-by star. We think the planet has the right conditions for life.

The remote sensing technology we're using has been working closer to home for years. Around the world, remote sensing has been used to help us with environment, climate, and weather management -- not days in advance, but months, years, and perhaps eventually, decades.

NASA's Earth Science program has been working in partnership with the private sector and other governments locally, regionally, and globally to predict regional rainfall and drought on a seasonal, an inter-annual, and ultimately a multi-decade basis.

Developing countries have been using remote sensing and other technologies to help better manage the precious water resources and agricultural products. For instance, this technology allows people to measure the variability of a specific field and literally "farm by the yard." We know exactly which area needs more irrigation and which area needs less. We know where there is too much fertilizer and what part of the field isn't getting enough. Same with pest control. We know exactly what portions of the field are ready for harvest and what portions are not. Space-based technologies are leading to better yields and a better quality of life.

Ever since 1996, private sector investment in space has far surpassed government expenditure on space activities. In addition to the mature commercial industries related to telecommunications, remote sensing, and launch services, new space-based industries have emerged. The private sector has realized growth associated with Lunar, Martian, and Near-Earth Asteroid opportunities, such as manufacturing and mining of space resources, energy production and transmission, and tourism.

Two members of the previously mentioned international crew had started training in 2001 and flew on a vehicle called the Space Shuttle. Experiments were conducted on the Shuttle that used the unique vantage point of space to look back at and monitor the Earth. We've come a long way since then.

For more than three decades, the Space Shuttle had served as the primary means to conduct human missions in low Earth orbit. In addition to serving as a platform for on-orbit research, it was also instrumental in the construction of the International Space Station. Originally developed in the early 1970s, the Shuttle system required the dedicated attention of hundreds of civil servants and thousands of support contractors for operations and maintenance. The members of the Space Shuttle Team refined and improved this vital transportation system throughout its operational life. During its service lifetime, the Shuttle Team compiled an enviable record of accomplishments in low-Earth orbit research which proved so vital for the development of the International

Space Station. Additionally, they remained the benchmark for excellence for air and space safety for decades to come.

As technology advanced and after the International Space Station was assembled, the research and transportation requirements for the Shuttle were transferred to a totally reusable launch system. Launch vehicles that were developed both with and without government support were able to provide a higher degree of reliability and operability that contributed to dramatic reductions in the costs of access to space. With less money directed toward shuttle operations, NASA was able to direct more of its scarce resources towards long range research and development activities. Shuttle Team members were retrained for new cutting edge assignments or while many transitioned to support the growing resuable launch vehicle industry.

Thanks to knowledge gained from the Shuttle system and consistent Congressional support for improved launch systems, the interstellar probe that the crew launched from Mars requires only a few controllers during the entire cruise phase. That's because the probe is a thinking, intelligent, and self-tasking spacecraft. It is too far away for commands to be transmitted from a mission control facility on Earth. At the speed of light, it would take months, even years, to relay the simplest communications. Just think how communications that go tens of trillions of miles will revolutionize communications on Earth that only have to go thousands of miles! And think of how advances in solar propulsion might assist the emerging space-based power industry!

This spacecraft can also learn and adapt as it travels. It has real-time health monitoring capability because the decisions are being made by the spacecraft itself. It is self-diagnostic and self-repairing. In many ways it is just like the human body. It has sensors and actuators. It reacts to stimuli. And it has a distributed nervous system with intelligence that enables the spacecraft to react and adjust according to changing environments.

This very same kind of technology is on display in the vehicle that flew from Japan to Florida. The vehicle has what we call "smart skin." It senses excessive heating and air loads and adjusts the shape of the craft for flow control and separation and load alleviation to avoid possible damage. The vehicle itself can take corrective action to change the lift and optimize performance. That's critically important while traveling at Mach 7 and at an altitude over 30 kilometers.

Back to our international crew of astronauts -- to ensure that our exploration activities were enduring and not single mission events, NASA placed an emphasis on facilities at strategic locations in the Solar System. For example, the solar Libration Point 1, is a stable location out of the Earth's gravity well that provides an excellent, nearby, viewing platform for Sun-Earth observations, astrobiological observations, and could serve as a transportation node for refueling. Similar locations throughout the Solar System will serve as nodes able to support research, communications, transportation, and in-space operations. These facilities will enable the conduct of a wider range of research, development, and exploration opportunities.

The research scientists and engineers working at these outposts are also using "smart" technology. The crew are wearing body forming "smart" suits that incorporate biosensors to record and report on various health parameters. Biological sensors are also used to report directly to the spacecraft control systems on air and water quality. A special electronic nose gives real-time reporting of any potential chemical or biological contamination that could threaten the lives of the astronauts.

Because the flight back from the outpost takes a few months, and the mission itself lasts a few years, the spacecraft used by our astronauts must be an ecosystem in its own right. It has a life-support recycling system that uses algae and other small organisms to rid the air and water of impurities and to generate oxygen.

The spacecraft also has incredible new tools and the capacity for virtual surgery. So if one of the astronauts gets sick, they won't have to worry about being so far away. The

53-067 99 - 2

other members of the crew will simply practice in virtual space and then conduct the procedure themselves. Simply put, the spacecraft is self-sufficient.

Our crew returns safely. As soon as the official reception committee congratulates the astronauts, the Japanese Prime Minister is informed that he is needed back home for an emergency meeting. Even with a two and a half hour flight, the Prime Minister will not be able to get to the meeting in time. However, like the heavy fog that existed when the Prime Minister arrived, this is not an insurmountable problem like it was back at the turn of the century.

The Prime Minister's seat on the vehicle becomes a virtual office. Advances in information technology and networking capacity allow for total immersion virtual presence. There is sight and sound and touch, and if you need it, even smell. The Prime Minister is briefed by his staff at home in a virtual environment. Then, while still on the vehicle, he participates in the full Cabinet meeting.

The other members of the Cabinet are projected in such clarity, it is like being there. You can't tell the difference. It is seamless. And it is only fitting, because this same virtual technology was used to simulate the entire development life cycle of the flight vehicle before anyone cut a single piece of hardware. Geographically distributed teams were able to be linked and development progress could follow the sun, making significant cuts in development cycle time and cost. It was used to simulate the harmonics of the entire factory where the vehicle was built before pouring any concrete or putting up any walls.

This virtual technology was incorporated into the training regime for astronauts as they prepared for their mission. It also enabled millions of people on Earth to join in the adventure of exploration throughout the journey.

The significance of these strategic goals that I have proposed here is about more than sending astronauts to outposts beyond Earth orbit and bringing them back. It's about the United States and other people and other countries working together to erase the distinction between air and space travel. It's about exploring new worlds in space and creating a new world here on Earth.

Opening the space frontier cannot depend solely on incremental evolutionary changes to the systems that enabled past achievements. We must also look to bold, innovative and high risk revolutionary leaps forward. What I have described here today is less science fiction than the notion of sending a human to the Moon was back on NASA's first day of operation in 1958. At NASA we are already conducting research on each of the technologies I have just described. And as exciting as the possibilities are to imagine, the most startling developments will no doubt occur in areas we have not even thought of yet.

Surgeons at Stanford University can now do reconstructive facial surgery on a three dimensional virtual light table and show the patient what they will be doing before they start.

Next year, a reusable launch vehicle will take off from the Mojave Desert in southern California and land at Malstrom Air Force Base in Montana -- 1,000 miles in under 25 minutes. It will hit Mach 15, yet the pilot will be on the ground in a virtual cockpit.

We are working on combustion experiments at 2000 degrees Centigrade that will have a profound impact on energy usage throughout the world.

The solutions to all these technology challenges are not yet apparent. Most are high risk and require long lead times for application. But all offer opportunities for American industries to maintain a competitive advantage in the global market place and unbelievable possibilities for future generations to enjoy a future of security and prosperity.

There will always be those who will tell us why we should not attempt to achieve this vision. We have heard the arguments before: It can't be done. It's too risky. It's too expensive. NASA is too much a bureaucracy to achieve such an ambitious agenda. There are higher priorities that need to be taken care of in the near-term.

I couldn't disagree more. In response to the Government Performance and Results Act, we have spent the last five years at NASA engaged in significant management reform and reinvention activities that will lay the foundation for new ways of doing business in the future. We have implemented procurement reforms to ease the burden of red tape and complex regulations. We have transferred direct program management responsibility from Headquarters to team members in the field and have established "Centers of Excellence" to foster and maintain world class capabilities at our research centers. We have increased opportunities for innovative partnerships that feature the sharing of cost and risk between the government and private sector. We are investigating the best methods to transfer functions that are not inherently governmental to the private sector. We have begun to incorporate data buys from the private sector to meet our science and technology mission goals and operational requirements. It is my hope that within ten years, NASA will have transferred all low Earth orbit operations and infrastructure to the private sector. We will then be able to focus our human and financial resources on pushing the frontiers of science and advancing technology.

We have initiated cooperative agreements with the Department of Defense, the National Institutes of Health, the Department of Energy, the National Oceanographic and Atmospheric Administration, and other Federal research and development entities. These agreements help to improve the returns from the Federal investment in research and development and identifies additional savings by eliminating duplication.

At the age of 40, there are those who declare us to be middle-aged and boring. I believe, however, that our middle-age milestone provides an excellent launch pad for superior achievements yet to come. The excitement generated during our youthful sprints has made way for sustained long-term marathons in the future. We are learning from our mistakes and we are taking corrective action to improve future missions.

You asked what policies and budget priorities would be needed to ensure that the vision I have presented here today will be realized. We believe we already have an excellent start. The Administration's National Space Policy and the Goals for Aeronautics Research and Technology, and Congressional Space Act. These documents provide direction and guidance to build the foundation for the vision I have described. The Agency's Strategic Plan defines how we intend to implement the near-term policy direction and guidance that is articulated in these documents. The Civil Space Guidelines section of the Space Policy specifies the research and development activities NASA is to lead and the priorities to follow. While the policy does not direct specific long-term destination goals, we believe it provides for modest investments in long-lead, high risk technologies for missions we will pursue once we have completed our commitments for the International Space Station.

In addition to stable policy guidance from the Administration, we will also require the sustained, bi-partisan advocacy that has characterized Congressional support for NASA for the past four decades. To earn this support, we intend to continue to do what we say we're going to do and honor our commitments to the Administration and Congress. We also intend to look for new ways to ensure that Congress plays a key role in the development of goals that will bring the highest payback to the taxpayers.

In the heat of the Cold War, NASA was created to demonstrate that the values and freedom of democracy was a superior means to accomplish bold and noble challenges. After four decades of achievement, there should be no doubt that NASA has succeeded in fulfilling the mandates from Congress as expressed in the Space Act. We have demonstrated capabilities and expertise that are the envy of the world and a model for other spacefaring nations.

As we enter the New Millennium and define new challenges for our aeronautics and space aspirations, the NASA Team is committed to returning high value for the public's investment. We will continue to set challenging goals and hold ourselves to even higher standards of excellence.

I would like to end with the words of a noted aerospace pioneer who devoted a good deal of time to the question of how humanity could benefit from the exploration and exploitation of space resources and capabilities. Speaking to a gathering of navigation engineers shortly after NASA's 10th anniversary, Dr. Krafft Ehricke of North American Rockwell outlined his vision for the future. In closing his remarks, Dr. Ehricke suggested that "Perhaps, as we place the extraterrestrial domain into the service of all people, we may be permitted to hope for the greatest benefit of all: that the ugly, the bigoted, the hateful, the cheapness of opportunism and all else that is small, narrow, contemptible and repulsive becomes more apparent and far less tolerable from the vantage point of the stars than it ever was from the perspective of the mudhole. After all, should we not take a cue from the fact that since the beginning, we have always placed our dreams and aspirations among the stars?"

We at NASA anticipate the future with great expectations. To bring the domain of space into the service of humanity. To do our part to realize the dreams and aspirations of a great nation in the decades to come.

Chairman ROHRABACHER. Thank you very much, Mr. Goldin.

Next, we have Howard McCurdy who is a Professor of Public Affairs at American University and a thoughtful author and commentator on NASA and space policy. He's currently the Charles A. Lindbergh Chair of the Air and Space History at the Smithsonian Institution, and you may begin. Thank you very much, Mr. McCurdy.

TESTIMONY OF HOWARD McCURDY, PROFESSOR OF PUBLIC ADMINISTRATION, AMERICAN UNIVERSITY

Mr. MCCURDY. Thank you. Mr. Chairman, members of the Committee, thank you for the opportunity to come here today. I'm glad to see so many of the members of the Committee here.

I brought a copy with me—if I could get one of the members of the staff to come up, and I'll hand it to you—this is a copy of the March 22, 1952, issue of Colliers Magazine. It is one of the visionary tracts that kicked off the Space Age over 40 years ago. You can see what it says on the top: "Humans will Conquer Space Soon," and right below it is a picture of a winged Space Shuttle advancing into low-Earth orbit. It's amazing to think how much of the vision that was formulated in the 1950's that led to the creation of NASA has been accomplished, and, of course, how much more still needs to be realized.

When we created NASA in 1958 in order to win the space race, the primary difficulty that people faced was concentrating what little rocket and satellite technology we had into a single agency so that we could win the space race. And the conditions that we face today are almost the opposite of the that. Due to the successes of the past 40 years, we now face the situation of diversification where many industries and many agencies are going to be participating in space, of which NASA is one, and I think in my oral testimony, what I'd like to do today is concentrate on what I think made NASA successful in the past and what NASA needs to do, especially in administrative terms, to be successful in the future.

Before that, let me just briefly summarize what kinds of organizations, government organizations in particular, are going to be needed for us to advance into the space frontier in the next 40 years. We're going to need some sort of a government-coordinating council that can make sure that the policies of the U.S. Government do not work at cross purposes, something along the lines of the National Space Council which I personally recommend be reestablished.

We're going to need some sort of government institution that can stimulate private investment in space. Think, for example, of a space bank organized along the lines of the federal housing authority which was created in order to spur housing development in the United States. Economists think that this might be an effective way of stimulating private investment in space, if there was some government mechanism that would guarantee or insure financing loans for private enterprise.

We're going to need the equivalent of a core of space engineers, perhaps modeled on the Army Corps of Engineers, in which the government, with its own financing, finances the infrastructure of space and needs a small group of people who are capable of monitoring the contracts to create that infrastructure which will be done by private industry. I think NASA is capable of doing this, but this is not exactly part of NASA's culture. The Army Corps of Engineers is an excellent example here on Earth of how that has been done.

And, of course, we also need an agency that can engage in what NASA does best, which is NASA itself. That is the process of research and development, the process of discovery, and also making sure that what we learn in space is applied back here on the Earth.

What I'd like to do is summarize very briefly what made NASA successful in the beginning and what would make NASA successful in the future. When NASA was formed, it had tremendous in-house technical capability. People like Wernher von Braun, brought over from the Army Ballistic Missile Program, Max Faget, who worked for the National Aeronautics Advisory—the National Advisory Committee on Aeronautics, were brought in to form this technical corps of NASA in order to assemble the most talented people we could find in the world to work on rocketry, satellites and spacecraft.

It began with this tremendous technical base, but it also began with a lot of disasters and failures, and it's recognized now that those disasters and failures were a result of not having the proper management systems in place. So, beginning at about 1963, NASA imported from the Air Force ICBM Program a number of officers and industrialists, such as George Mueller and General Samuel Phillips, who brought in a system of project management that was revolutionary for its time and today is copied worldwide. One of NASA's most important spinoffs are the methods of management that it developed in the 1960's to conduct programs like the Ranger probes to the moon, and, of course, Project Apollo. Joined together, that technical capability and that systems management created the NASA that we have today and made it strong.

And then in the 1970's and 1980's, some of that capability was lost. It's been regained in the 1990's, but if we're going to have a strong NASA for the future, we have to make sure that NASA has very strong in-house technical capabilities, sometimes built on the ability to do its own work. There's no reason in the world why we can't have a strong industry and a strong NASA both practicing the same principles, both with their same technical base of capability, and there has to be very strong project management teams.

Look at the recent success of the Pathfinder team at the Jet Propulsion Laboratory that put together that wonderful Mars mission [Pathfinder]: 230 dedicated individuals working overtime most of the time to conquer the problems of making that spacecraft work. They had the technical discretion, they had the ability to understand what was necessary to carry out that project, and they could do it with a fairly free hand. They also, I might add, built a considerable amount of that spacecraft in house which gave them the technical expertise they needed for the base.

I'd also just like to point out that NASA works best when it doesn't do routine operations. NASA's always been an institution where once it learned how to do something it spun it off and moved on to something new. And also NASA is a good organization when it's rewarded for taking risks.

Let me just add one other thing that perhaps the Committee would like to consider. I'm enamored by the idea of lump-sum appropriations. I think if you're going to strengthen project management, one of the best ways to do that is to tell the project team, "This is how much money you've got. Report back to us at periodic intervals on whether or not you're meeting your financial milestones. If you exceed those milestones by 15 percent, the project is canceled as quickly as we can." But give them a lump-sum appropriation so that the project team understands the best possible way to spend the money, so it can invest as much as possible in hardware. That might consist of a line of credit or some other form that doesn't have to be exactly a lump-sum appropriation. I think it worked very well for the Challenger replacement, and I think it would be essential for assuring that NASA is a vital and strong institution in the future.

We invested tremendously in NASA when it was created 40 years ago, and I think as we invest in all of the institutions that are going to carry us into space for the next 40 years, we have to make sure that we have a strong NASA, as well, during that period. Thank you very much.

[The prepared statement and attachments of Mr. McCurdy follow:]

NASA at 40:
What kind of space agency does American need for the 21st century?

Testimony of Dr. Howard E. McCurdy
Subcommittee on Space and Aeronautics
Committee on Science
U.S. House of Representatives
October 1, 1998

My name is Dr. Howard E. McCurdy. I am a professor of public affairs at American University in Washington, D.C., where I specialize in public administration and space policy. This year I am serving as the Charles A. Lindbergh Chair in Aerospace History at the Smithsonian Institution. I have published four books on the NASA space program, including a history of NASA's changing organizational culture, and am currently working on a study of the "faster, better, cheaper" initiative.

For the purpose of this hearing, you have asked me to comment on U.S. strategic goals in space, the appropriate roles of the federal government, and how one would design a civil space agency to pursue those goals. I will concentrate on the type of agency that is needed for the next forty years, based on the history of what has worked best over the past forty.

Strategic goals in space: The original vision that launched the era of space exploration forty years ago is essentially unchanged. We still seek to explore space with human and robotic missions and make discoveries that change the way we view the Earth and our place in the cosmos. We still want to construct large structures that will make that possible – space stations, orbiting telescopes, lunar bases, Mars colonies. We still have a vision of cheap, easy to fly spacecraft. We still want to use space to alter the way in which we live here on the Earth.

After forty years, we are halfway there. Half of our vision is coming true. We continue to make fantastic discoveries about our Moon, the solar system, and nearby stars. We use space, particularly the realm out to geosynchronous orbit, in ways that continue to transform technology and economics here on Earth.

Still, we still have far to go. Our efforts to develop cheap, reusable spacecraft that will reduce the cost of space flight "by a factor of ten" have been disappointing. We are far behind in our original schedule to assemble the first large modular structure in space, an International Space Station. We have achieved far more than the public believed we could fifty years ago, when only 15 percent of the people responding to a 1949 Gallup public opinion poll agreed that humans by the millenium would reach the Moon. But we have not accomplished as much as many visionaries believed we would.

The conditions that guide our ventures into space are considerably different than those that led to the creation of the National Aeronautics and Space Administration forty years ago. In 1957, when the Soviet Union launched Sputnik 1 and 2, the U.S. space effort was fragmented and disorganized. Nearly all of our rocket and satellite research was concentrated in the military. Congress moved to consolidate as much technical capability as the U.S. could spare into a single civil agency that could win the "race" into space. Due to congressional persistence and the decision to go to the Moon, NASA became the agency that it is today -- a single multi-center organization capable of carrying out complex missions (sometimes rapidly) at very high levels of reliability.

Conditions have changed significantly since then. The area of "near space" has become an extension of what we routinely do on Earth. Just as no single government agency implements all of the public policies that affect our land and sea mass, no one agency conducts everything we do in the space near Earth.

Industrial capability for operating in space is far more robust than it was forty years ago, when our civil space talent could be assembled in a single government organization.

The guiding principle for the government's civil space program has changed. To be sure, the U.S. continues to derive technological, economic, and scientific benefits from being the world's leading "space power." But the space race is over, and with it, the guiding principle for the creation of NASA. We no longer need spectacular programs developed quickly at any cost. We need effective programs developed quickly at low cost. Unless we can reduce the cost of space operations, we are going to be stuck here on Earth for a very long time.

These are the most important strategic goals for the next forty years. We need to develop the infrastructure of space, beginning with the International Space Station and continuing through space habitats and large communication networks. We need to develop cheap, reusable spacecraft. We need to expand the involvement of industry and other government agencies in space. We need to significantly reduce the cost of space operations. And we need to continue a robust program of space exploration, research, and development.

Roles of the federal government: NASA is capable of meeting some but not all of these challenges. Many government agencies will manage space as we move further into the void, and not all of them will look like NASA. In the beginning of the space age, we needed to consolidate our civil space expertise in one organization. Now that the use of space has matured, we need coordinating councils, national banks, engineering corps, and research and development agencies.

We need one governmental body that can coordinate the space activities of the U.S. government and ensure that the policies of individual agencies do not work at cross-purposes. Personally, I favor reestablishment of the National Space Council, although an agency modeled along the lines of the Federal Aviation Administration could perform this work as well.

We need better methods of promoting private entrepreneurs in space. In my view, the most promising approach for stimulating private investment in space is a "space bank" that could provide venture capital to space entrepreneurs, perhaps in the form of government guaranteed or low interest loans. The agency that does this will look more like the Federal Housing Authority than the old NASA.

To develop the infrastructure of space, we need something like a Corps of Space Engineers. perhaps within NASA, in which a few very qualified civil servants monitor the performance of private companies operating under government contracts. The corps need not "penetrate" the work of contractors in the same way that NASA did in the past, but it does need to assure accountability where public funds are involved. The obvious model for such activity is the Army Corps of Engineers or the various port and stadium authorities in the U.S.

What does that leave as NASA's central role? In my opinion, NASA should do what it does best. It should innovate and explore, conducting research and organizing missions that lead the rest of us into space. We need a government agency that conducts research in support of private industry and the government itself. That agency should design and test experimental spacecraft. And it should continue to probe the unknown with expeditions both human and robotic.

This brings me to the point where I hope my expertise is useful. To be a successful research and development agency, NASA needs to possess special organizational strengths. In our rush to contract out space activities and encourage private investment in space, we often lose sight of what those characteristics are. I should add that those strengths are equally required for industrial activities on the high frontier. The history of space flight over the past 40 years shows in a dramatic way what happens when those principles are ignored.

Redesigning NASA: NASA began its work forty years ago with plenty of in-house technical capability. To form NASA, government leaders assembled five government research laboratories, transferred in the German rocket team and their allied group of American engineers led by Wernher von Braun, and created two new centers devoted to spacecraft and satellite development. We assembled the strongest concentration of people with hands-on experience in rocketry and spacecraft development in the world.

In the beginning, NASA's technical capability was not matched by its organizational skill. Said one early leader, "NASA had considerable technical depth, but almost no program management experience." As a consequence, mishaps plagued the early space flight effort. The first six Ranger probes to the Moon failed. An aerospace trade journal characterized Project Mercury in the year before Alan Shepard's successful 1961 suborbital flight as a "flaming tragedy." The Gemini program was plagued by cost-overruns and test wreckage.

To correct these problems, NASA executives turned to military and industry leaders from the defense department's intercontinental ballistic missiles (ICBM) development program. These people had invented systems management techniques capable of controlling very large technical undertakings. Those techniques were imposed on the there-to-fore decentralized and fragmented field centers that made up the original NASA.

The organization that arose was hugely successful. It dispatched the Voyager spacecraft on a Grand Tour of the solar system, placed Viking landers on the planet Mars, and sent American astronauts to the Moon and back. The systems management techniques NASA perfected were copied around the world. NASA's early accomplishments were based upon traditions and capabilities that can be summarized as follows:

1. Tremendous in-house technical capability. NASA's research labs and von Braun's rocket team contained people with extensive hands-on experience.
2. Very strong, centralized systems management. Single centers of authority with enormous technical discretion led NASA's early space flight efforts.
3. Substantial risk-taking. NASA employees were allowed to take plenty of risks but told not to let the overall mission fail. Failure was a primary means by which flight engineers learned how new technologies worked.
4. A frontier mentality. As soon as NASA and its contractors completed a mission, they moved on to something new. NASA kept people sharp by letting them work on

new challenges. In the early years, there was not much tolerance for doing the same thing over and over again.

All of these traditions began to weaken as NASA aged. NASA lost much of its in-house technical capability to contracting out. Centralized systems management techniques were replaced by procedures that divided centers of authority, such as the much weaker "lead center" approach. "Work packages" on projects like the International Space Station were distributed in ways designed more to protect field centers than to assure program success. The agency spent a disproportionate amount of time operating systems like the space shuttle that had been declared "fully operational."

As the old traditions weakened, NASA's performance suffered. The failure of the space shuttle to meet its original cost goals and the difficulties NASA has encountered in designing the International Space Station can be traced in part to the erosion of capabilities from the early space program.

Policies and priorities: To meet the challenges of the next 40 years, NASA needs to renew its capabilities. Technical innovation and cost reduction will not occur unless this is done. Top officials need to strengthen the technical capacity and systems management techniques that characterized NASA during the first decade of space flight. Some of this has already been accomplished under Administrator Dan Goldin, but much more needs to be done.

To assist in this effort, the Space and Aeronautics subcommittee could use its oversight powers to promote NASA's management capabilities. Perhaps it could hold hearings on the management of the future space program, as it did 34 years ago when the first great era of space exploration began. (See House Committee on Science and Astronautics, Subcommittee on Manned Space Flight, 1965 NASA Authorization, 88th Cong., 2nd sess., pt. 2, 1964.)

NASA needs more in-house technical capability. It needs to hire more bright, young engineers. It needs methods to promote a better flow of talent between government and industry. Congress might consider the creation of positions exempt from civil service regulations, as it did in the original 1958 legislation establishing NASA.

NASA needs more hands-on projects like Mars Pathfinder. That spacecraft and its Sojourner rover were built in-house at the Jet Propulsion Laboratory. In-house projects strengthen technical capability, and can lead to dramatic break-throughs in the drive toward "faster, better, cheaper" space initiatives.

I know that the call for in-house technical capability flies in the face of current trends toward privatization and contracting out. There is no reason, however, why NASA cannot do both. It can oversee technologically mature activities that which are financed by the federal government but conducted primarily by private industry, while at the same time conducting research and development using its own technical base. When a technology is immature, NASA should use its full capabilities to develop it. Once the technology becomes mature, NASA should step back and let industry take over. Since each approach embodies such a different culture, it might be advisable to assign industry oversight to a separate field center. But both can be done.

For its basic exploration and development activities, NASA needs very strong project management teams. Those teams need not be large, but they must be powerful. A single point of authority remains the one best way to carry out new space ventures. Especially to reduce cost, NASA needs small project management teams free to conduct their work in the most technically effective way. The project teams, with a few employees as possible, need a great deal of technical discretion. Team members should have the authority to decide how much work they are going to do in-house, what activities they are going to contract out, and what support they need from other centers.

A number of forces currently conspire to divide project authority. The desires for multi-center participation, contracting out, and international participation all have the potential to erode project authority.

In general, multi-center programs create more problems than they solve. Project Apollo had to be a multi-center undertaking because NASA's spacecraft talent and rocket expertise were located at two separate field centers. NASA compensated for this division by creating a very strong systems engineering and integration team under General Samuel C. Phillips here in Washington. D.C. The Phillips team used configuration control and other systems management techniques to coordinate the work of the centers involved. Also, as the engineers like to say, the interfaces between the rocket and the spacecraft were "very clean" (the spacecraft sat on top of the rocket). And Phillips had enough money to run his systems management organization. It was a costly solution, but it worked.

What was necessary for Apollo too often becomes convenient for protecting field centers. Wherever possible, Congress should insist that total project authority for missions large and small be vested at a single NASA field center. The practice adopted for the International Space Station, in which work packages were distributed to NASA field centers before the creation of a project management team, should be discouraged. As a guiding principle, a strong project team should always be created first.

International cooperation tends to create the same complications as multi-center programs. Sometimes international cooperation is desirable for reasons of its own. In those cases, the complications can be mitigated somewhat by allowing project management teams to deal directly with service or hardware providers in the participating countries. Persons above the project team can negotiate the scope of international participation, but any attempt to resolve technical details at higher levels will complicate project management.

Congress should experiment with the granting of lump sum lines of credit equal to the entire estimated cost of future space missions. If NASA managers estimate that the cost of a particular project is X, they would receive a line of credit for X with the proviso that any cost growth above 15 percent (based on milestones along the way) will result in project cancellation. In principle, this would encourage project managers to maximize the amount of money invested in flight hardware and minimize the amount of money spent on personnel and overhead. This approach seemed to work fairly well for the Challenger replacement and is being applied in part (through fixed assured costs) for NASA's "faster, better, cheaper" initiatives.

NASA went though a period in the 1970s and 1980s when it lost many of the strengths that characterized its early work. Making sure that the revitalization of NASA continues in appropriate ways will help assure that our strategic goals are fully realized during the next forty years.

HOWARD E. McCURDY

5001 Sentinel Drive, Unit 14
Bethesda, Maryland 20816

Telephone: 202/885-6236
mccurdy@american.edu

Books published

Space and the American Imagination. Washington: Smithsonian Institution Press, 1997.

Space Flight and the Myth of Presidential Leadership, with Roger Launius (coeditor). Urbana: University of Illinois Press, 1997.

Inside NASA: High Technology and Organizational Change in the U.S. Space Program. Baltimore: Johns Hopkins University Press, 1993. (Winner of the 1994 Henry Adams prize from the Society for History in the Federal Government for the best book on the history of the federal government.)

The Space Station Decision: Incremental Politics and Technological Choice. Baltimore: Johns Hopkins University Press, 1990. (Selected by *Choice* magazine as one of the "Outstanding Academic Books" in 1991.)

Public Administration: A Bibliographic Guide to the Literature. New York: Marcel Dekker, 1986.

Public Administration: A Synthesis. Menlo Park, CA: Benjamin/Cummings, 1977.

Other Publications (selected)

"Fiction and Imagination: How They Affect Public Administration," *Public Administration Review* 55 (November-December 1995) 499-506.

"Organizing for Space: The Popular Culture of the Cold War," in Roger Launius, ed., *Organizing for the Use of Space*, AAS History Series, vol. 18. San Diego: American Astronautical Society, 1995.

"The Costing Models of the Early 1970s and the Launching of the Space Shuttle Program," in Emmanuel Chadeau, ed., *L'Ambition Technologique: Naissance d'Ariane.* Paris: Istitut d'Histoire de l'Industrie, 1995.

"Public Policy and Popular Imagination," in Charles T. Goodsell and Nancy Murray (eds.), *Public Administration Illuminated and Inspired by the Arts.* Westport, CT: Praeger, 1995, pp. 178-88.

"The Cost of Space Flight," *Space Policy* 10 (November 1994) 277-89.

"Where Space Begins: Politics, Presidents, and Congress," in Martin J. Collins and Sylvia K. Kraemer, eds., *Space: Discovery and Exploration.* Washington: The Smithsonian Institution, 1993.

"NASA's Organizational Culture," *Public Administration Review* 52 (March/April 1992) 189-192.

"Organizational Decline: NASA and the Life Cycle of Bureaus," *Public Administration Review* 51 (July/August 1991) 308-315.

"The Space Station Decision: Politics, Bureaucracy, and the Making of Public Policy," in Martin Collins and Sylvia Fries, eds., *A Spacefaring Nation: Perspectives on American Flight and Space History*, Washington: Smithsonian Institution Press, 1991), 9-28.

"The Decay of NASA's Technical Culture," *Space Policy*. (November 1989) 301-310.

"The Decision to Build the Space Station," *Space Policy* (November 1988) 297-306.

The Rivers of Mars, by Piers Bizony. Reviewed in *Space Times* (November-December 1997) 21.

Pale Blue Dot: A Vision of the Human Future in Space, by Carl Sagan. Reviewed in *Nature* 375 (May 25, 1995) 287.

"Justifying Space Exploration: Lessons from History," *Space Times* (March-April 1994) 10-12.

"Piece by Piece: Incrementalism is the Rule of Thumb for Today's Disjointed Space Program," *Ad Astra* (The Magazine of the National Space Society) 1 (February 1989) 24-28.

"The Graying of Space: Public Opinion Now Appears High for Space But Will an Aging NASA Be Able to Ride This Tide to Mars -- or Beyond?" *Space World* 10 (October 1988) 10-12.

Education

Ph.D., Cornell University (Graduate School of Business and Public Administration), 1969.

MA, University of Washington, 1965; BA, University of Washington, 1962.

Professional Experience

Professor of Public Affairs (tenured); The American University, 1968 to present.

Charles A. Lindbergh Chair in Aerospace History, Smithsonian Institution, 1998-99.

Visiting Professor, Graduate School of Public Affairs, University of Washington, 1995-96.

Visiting Professor, Public Administration, Lewis and Clark College, Summer 1983.

Fulbright-Hays Lecturer, Department of Political and Administrative Studies, University of Zambia, 1978.

Management Analyst, U.S. Bureau of the Budget, 1966 and 1967

Legislative Analyst, Washington State Legislature, 1963 and 1965

Chairman ROHRABACHER. Thank you very much. That was very enlightening, and we'll have some questions for you afterward.

And, I'm delighted today that we have with us Dr. Galloway, who is a true pioneer in America's space program. As a Research Analyst who started at the predecessor office of our Congressional Research Service; she actually started there 6 years before I was born——

[Laughter.]

Chairman ROHRABACHER. And Dr. Galloway made unique and critical contributions to the Congress' early work on space, including helping to draft the NASA Act of 1958, and she has continued to actively participate in the space law community, particularly international space law. So, I'm delighted that Mr. Gordon's staff recommended that she be a witness today, and we're very, very happy to have you with us. So, you may proceed.

TESTIMONY OF EILENE GALLOWAY, HONORARY DIRECTOR, INTERNATIONAL INSTITUTE FOR SPACE LAW

Ms. GALLOWAY. Yes, Mr. Chairman and members of the Committee. Thank you for inviting me to give my views on NASA at 40: What kind of space agency America needs for the next century. With your permission, I would like to submit for the record a paper I wrote last year on the occasion of the 40th anniversary of Sputnik entitled, "Organizing the United States Government for Outer Space 1957 to 1958." It is appended to these remarks and reviews how Congress responded to the Sputnik launch including the role of creating NASA. The paper may be useful as you assess NASA's first 40 years and deliberate on what changes may be needed as the Agency moves into the next century.

Chairman ROHRABACHER. So ordered.

Ms. GALLOWAY. Yes. On NASA's 40th birthday, we can celebrate 4 decades of peace in outer space and freedom from space wars. I think that's the most important thing. This is a tribute to the policies and programs we adopted at the beginning of the space age, and it is a remarkable achievement considering that when the Soviet Union suddenly launched Sputnik on October 4, 1957, the reaction was fear that outer space had become an arena for warfare, and Sputnik was perceived as a crisis for U.S. national defense.

Hearings were immediately begun by the Senate Preparedness Investigating Subcommittee and led to the conclusion that the United States must make a tremendous effort to achieve preeminence in space. Space scientists and engineers testified that rocket technology could be used not only for weapons but also for beneficial civilian purposes. Then the definition of our problem expanded from fear of orbiting weapons of mass destruction to hope for producing benefits for mankind. We were faced with a choice between peace and war.

Congress created new committees for the legislative task of organizing the government for space defense and civilian applications. The House of Representatives established a Select Committee on Astronautics and Space Exploration. The Senate created the special Committee on Space and Astronautics. Both Committees were chaired by the majority leaders of the House and the Senate, Congressman McCormick of the House and Senator Lyndon Johnson.

A consensus developed naturally among technical, political, national, and international decisionmakers, because they agreed on the objective: to maintain outer space as a dependable, orderly place for space exploration and peaceful civilian uses. The intent of the framers of the NASA Act was that the Agency would occupy a high place in the Executive Branch. The President would be involved, and NASA would be a leader. NASA was restricted to operations for research purposes, a provision that can diminish its role when it's interpreted too strictly.

I think there should be more flexibility for operations in astronautics for unusual circumstances. More emphasis should be placed on space exploration in the future. NASA has produced an outstanding civilian space program, especially in international cooperation with nations and groups of nations, and this function should continue to be emphasized.

NASA's role now and in the future is not confined to provisions in the NASA Act, and in planning for the future, we must consider that the Agency is also bound by the national space policy as determined by the White House National Science and Technology Council, by space legislation for other agencies, as well as international treaties. The United States Constitution provides that ratified treaties become the supreme law of the land. So, space law treaties as well as others related to environmental affairs contain provisions that NASA must observe. Such treaty provisions include registration of space vehicles, special treatment for astronauts and space objects, authorization and continuous supervision of governmental and non-governmental entities, and liability for damage.

We are now at a critical time for making decisions for the future, and they cannot all be done via the NASA Act. Private industry is now outpacing the government in space development and services, especially in global communications. This will create new relationships between government and the private sector—regulation and space activities who have been by sovereign States in terms of national security, safety, dependability, efficiency, and avoidance of conflicts.

There are questions about the consequences of the growing number of launchings, of the selection of orbits for specific functions, technology transfer, insurance, government liability for damage, and location of space ports. We should have no more regulation than is necessary for specific purposes, but it is essential to decide definitely what should be regulated, how, and by whom. Uncontrolled deregulation would result in chaotic situations for programs that demand precise scheduling. In fairness to all, we should devise a regulatory system that is practicable and well-understood.

Congress now has the opportunity to take the lead as it did in 1958 to working out a solution to the new problems arising in relations between the government and the private sector on the regulation of space activities.

[The prepared statement and attachments of Ms. Galloway follow:]

NASA AT 40:
WHAT KIND OF SPACE AGENCY DOES AMERICA NEED FOR THE 21ST CENTURY?

Testimony of Dr. Eilene Galloway
Honorary Director, International Institute of Space Law
Trustee Emeritus, International Academy of Astronautics

Before the
Subcommittee on Space and Aeronautics
House Committee on Science

October 1, 1998

Testimony of Dr. Eilene Galloway
October 1, 1998

Mr. Chairman, members of the subcommittee, thank you for inviting me to present my views on "NASA at 40: What Kind of Space Agency Does America Need for the 21st Century?" With your permission, I would like to submit for the record a paper I wrote last year on the occasion of the 40th anniversary of Sputnik entitled "Organizing the United States Government for Outer Space: 1958-1959." It is appended to these remarks, and reviews how Congress responded to the Sputnik launch, including its role in creating NASA. The paper may be useful as you assess NASA's first 40 years, and deliberate on what changes may be needed as the agency moves into the 21st century.

On NASA's 40th birthday we can celebrate four decades of peace and freedom from space wars. The policies we adopted at the beginning of the space age have been implemented by programs that preempted outer space for peaceful benefits for all mankind. This is an amazing achievement when we recall that on October 4, 1957 when the Soviet Union rocketed the Sputnik into outer space, the reaction was fear that this new environment would became an arena for warfare. The dramatic event was perceived as a crisis for U.S. national defense.

Both the United States and the Soviet Union had outer space projects in connection with the International Geophysical Year (July 1, 1957-December 31, 1958) when 67 nations cooperated in worldwide scientific investigations of the Earth, oceans, atmosphere and outer space. The United States was working on the Vanguard satellite which weighed 3.25 pounds. The Sputnik weighed 184 pounds, and the perception of a military threat was increased when the second Soviet satellite weighing 1,120 pounds was launched on November 3, 1957.

The definition of the problem was expanded when scientists and engineers explained that these powerful rockets could be used not only for weapons but also for benefits to global space communications, weather forecasting, and information from remote sensing that could protect the Earth's environment. This gave us two strong motives for future planning: Hope and Fear. Our choice was between peace and war.

At that time there were three main centers for decisionmaking that could lead to action: organized scientists and engineers who were experts in space technology and with experience in international cooperation; nation states that could formulate policies and fund programs; and the United Nations which became the forum for the negotiation and formulation of international space law.

The reason these three groups naturally merged into consensus is that they agreed on the objective: to prevent orbiting weapons of mass destruction and ensure that outer space be maintained for peaceful purposes. They accepted the fact that space activities are naturally international as satellites quickly orbit the Earth. Tracking stations were needed throughout the world and satellites require some regulation. Nations were even willing to renounce claims of sovereignty in outer space and over celestial bodies in order to eliminate a potential source of international conflict.

Congress played an outstanding leadership role. Because of the connection between satellites and national defense, the first hearings were held by the Senate Armed Services Committee's Preparedness Investigating Subcommittee, chaired by Senator Lyndon B. Johnson, the Majority Leader. The "Inquiry into Satellite and Missile Programs" began the month after Sputnik, on November 25, 1957, and continued into 1958. Expert witnesses were called from government, industry, and the scientific community to testify on the military and civilian uses of outer space. The Subcommittee concluded that it was essential for the United States to make a tremendous effort to ensure the nation's preeminence in defense and space exploration. The scope and variety of civilian uses were beyond the jurisdiction of the Preparedness Investigating Subcommittee, and on February 6, 1958 the Senate created the Special Committee on Space and Astronautics to consider how the government should be organized to achieve an outstanding space program.

Because of the urgency of the situation, and pending decisions on government organization, Congress passed interim laws to accelerate the satellite program. On February 12, 1958, the Secretary of Defense was authorized to engage in advanced research, and for one year develop space projects designated by the President (Public Law 85-325). Another law gave the Department of Defense $10 million in supplemental funds for space activities undertaken during the remainder of the fiscal year (Public Law 85-322, February 11, 1958). The Advanced Research Projects Agency was established in the Department of Defense to proceed with space development while Congress considered the future.

On March 5, 1958 the House of Representatives created the Select Committee on Astronautics and Space Exploration with the Majority Leader, Honorable John W. McCormack, as Chairman.

Congress had to consider not only the organization of the Executive Branch but also that of the legislature. It could be foreseen that space activities cut across the jurisdiction of so many committees that problems would be created in the referral of bills. In addition to Appropriations, bills on meteorology and reconnaissance satellites could concern the Committees on Armed Services, Foreign Relations, Commerce, Government Operations, and the Joint Committee on Atomic Energy. The space program was so important that a focal point in the House and Senate was needed for legislation.

It was also necessary to provide for the interest of all these committees in space technology, and the solution, adopted by the House and Senate, was to appoint the chairman and ranking minority members to the newly established space committees.

At that time I was National Defense Analyst in the Legislative Reference Service, now the Congressional Research Service, and had written a report published by the House and Senate on "Guided Missiles in Foreign Countries." Immediately after the Sputnik was orbited, Senator Richard B. Russell, chairman of the Senate Armed Services Committee, arranged with Senator Lyndon B. Johnson for my assistance with the space hearings of the Preparedness Investigating Subcommittee, an experience that proved invaluable when I was requested by Congressman McCormack and Senator Johnson to work through the legislative process that led to the NASA Act. Both House and Senate Committees were studying the issues involved in organization and functions before President Eisenhower sent his special message to the Congress on April 2, 1958 recommending "that aeronautical and space science activities sponsored by the United States be conducted under the direction of a civilian agency, except for those projects primarily associated with military requirements."

There was no difference of opinion between the Congressional committees and the President on using the National Advisory Committee for Aeronautics (NACA) as the nucleus for the new civilian space agency or that the Department of Defense be responsible for space applications required for national defense. However, modifications were made by the House Committee in broadening the scope of NASA, giving the Director more authority, changing the name from "agency" to "Administration" with an Administrator. More coordination was provided between NASA and the Department of Defense and Atomic Energy Commission, and the provision on international cooperation was strengthened. The Senate Committee introduced the coordinating concept of the National Aeronautics and Space Council as well as adding Section 205 on International Cooperation to the bill. Congressman McCormack introduced Concurrent Resolution 332 on June 2, 1958 on the peaceful exploration of outer space which passed the House on June 2, 1958 and the Senate on July 23, 1958. The purpose of these changes was to place NASA at a high level in the government, and ensure NASA's capability to carry out the declaration of policy and purpose as stated in Section 102 of the NASA Act: "The Congress hereby declares that it is the policy of the United States that activities in space should be devoted to peaceful purposes for the benefit of all mankind."

From the perspective of 40 years, there are several important subjects that deserve comment.

NASA was placed in a strong position within the Executive Branch and in relation to separate Congressional space committees. However, a strict interpretation of the policy declaration that space activities "shall be the responsibility of, and shall be directed by, a civilian agency exercising control over aeronautics and space activities" except for those in the Department of Defense, would lead to the conclusion that NASA does not have that much authority over other agencies with space-related functions. This did not deter NASA's development of space exploration and beneficial civilian uses at the highest level. However, the concept of NASA's basic role has been affected by a strict interpretation of the definition of "operations" in relation to research.

The definitions in Sec. 103 include "(B) the development, construction, testing, and operation for research purposes of aeronautics and space vehicles." This concept of delimiting operations to "research purposes" is evidently a carryover from the NACA where it was appropriate for relations between the government and the aviation industry, but it does not take account of the unique features of outer space and the fact that the government is the chief user of some satellite-derived information which cannot be obtained from the private sector. It might be efficient and economical on occasion to permit exceptions. It was easy to turn communication satellites over to the Communications Satellite Corporation because there was a thriving profitable industry in place to receive this technology. It was convenient to turn meteorological satellites over to the Department of Commerce because weather prediction was historically a government function. But when it came to turning remote sensing over to private industry which could not make a profit, the government was deprived of valuable information. It would seem that the mindset of reducing NASA to strict R&D without any flexibility for authority for operations in unusual circumstances is a feature worth examining for the future.

Coordination of basic space policies is another concept that bears watching. Congress recognized the need for coordination of space activities in the original NASA Act but the Executive Branch eliminated the institutions passed by Congress: the National Aeronautics and Space Council and the liaison committees. The intent of the framers of the legislation was that there be a small permanent staff which would develop an institutional memory to assist an advisory council on space matters that involved more than one agency. Inter-agency committees for specific problems as they arise is not the answer to this problem. We are in danger of falling back to our original position in 1958 when significant space matters can be referred to several agencies whose authorizations and appropriations are matched to different committees in the Congress. I do not know if the Office of Science and Technology Policy is staffed for this purpose.

NASA has an outstanding record in international space cooperation. Through bilateral and multilateral agreements, NASA has developed successful projects with most of the countries of the world. The Administration has carried out the declaration of policy and purpose with which the Act begins, that "The Congress hereby declares that it is the policy of the United States that activities in space should be devoted to peaceful purposes for the benefit of all mankind," and has cooperated with other nations and groups of nations in peaceful results. The U.S., through the Department of State and NASA, has exercised leadership in the United Nations in formulating international space law to guide nations in the conduct of their space activities.

We can recall that dramatic event in history when President Eisenhower sent a plane to Texas to fly Lyndon Johnson to the United Nations to speak on November 17, 1958 in support of the U.S. proposal to create the Ad Hoc Committee on the Peaceful Uses of Outer Space, thus demonstrating bipartisan support for the U.S. for international cooperation. Eighteen nations joined in this effort and a year later this Ad Hoc Committee became the permanent Committee on the Peaceful Uses of Outer Space which now has 61 member nations. This is the body where the U.S. has exercised leadership in drafting four space treaties which the U.S. has ratified.

The United States Constitution provides that ratified treaties become "the supreme law of the land" and planning for the future must take into account the following international space laws:

Treaty on Principles Governing the Activities of States in the Exploration and Use of Outer Space, including the Moon and Other Celestial Bodies. (October 10, 1967)

Agreement on the Rescue of Astronauts, the Return of Astronauts and the Return of Objects Launched into Outer Space. (April 22, 1968)

Convention on International Liability for Damage Caused by Space Objects. (March 29, 1972)

Convention on Registration of Objects Launched into Outer Space. (January 14, 1975)

These treaties are now on the agenda of the Legal Subcommittee of the United Nations Committee on Peaceful Uses of Outer Space for review.

We are now at a critical time for making decisions that affect relations between governments and the private sector. During the past 40 years governments have funded and developed space activities, but now it is estimated that private industry, national and international, are outpacing governments. In the next 5 years, the commercial market is likely to predominate. The UNISPACE III report on "Commercial Aspects of Space Exploration Including Spin-Off Benefits" reports that "The global space industry has become one of the largest industries in the world, with an estimated $77 billion in revenues and more than 800,000 people employed world-wide in 1996..." The 1997 outlook on the "State of the Space Industry" indicates "a growth in US$ millions by the year 2000 of 6,079 for government and 8,510 for commercial." (UNISPACE III Background Paper No. 7 A/CONF.184/BP/7, May 27, 1998, pp. 4,6.)

This situation calls for decisions on the regulation of space objects. Regulation has been controlled by nation states for reasons of safety, dependability, efficiency and avoidance of conflicts. We have the International Civil Aviation Organization as a model for regulating aviation. The International Telecommunication Union has authority over the geostationary orbit and allocation of frequencies. International regulation has been based on relations between nations — inter - national — but now global activities are emerging, especially in telecommunications, and may become outside the control of States. We will face difficulties in solving this problem because the decisionmaking groups that merged in the first days of the space age are no longer united, peace is taken for granted, there is no longer the fear of space wars, and we live at a time when the psychological atmosphere favors deregulation.

We should have no more regulation than is necessary for specific purposes. It is essential to decide what should be regulated, how, and by whom. It would be unacceptable to let trends develop and form patterns that become beyond control.

ORGANIZING THE UNITED STATES GOVERNMENT FOR OUTER SPACE: 1957-1958

By

DR. EILENE GALLOWAY

Honorary Director

International Institute of Space Law

RECONSIDERING SPUTNIK
FORTY YEARS SINCE THE SOVIET SPUTNIK

Symposium Sponsored by

NASA Office of Policy and Plans/History Office
National Air and Space Museum
George Washington University Space Policy Institute
Kennan Institute for Advanced Russian Studies

Smithsonian Institution

September 30-October 1, 1997

ORGANIZING THE UNITED STATES GOVERNMENT FOR OUTER SPACE: 1957-1958

By

EILENE GALLOWAY

Defining the Problem

The dramatic orbiting of the first Sputnik by the Soviet Union on October 4, 1957 was like a spark that ignited and speeded the process of developing the exploration and peaceful uses of outer space on a continuing and larger scale. People throughout the world were astonished by this phenomenal opening of outer space as a new environment, and surprised that the Soviet Union was first to accomplish this feat. But the news struck Capitol Hill like a thunderbolt because thrusting the 184-pound satellite into outer space was evidence of the capability of launching intercontinental ballistic missiles, and therefore instantly perceived as a crisis for U. S. national defense. This perception was reinforced on November 3, 1957 when a second Sputnik weighing 1,120 pounds began circling the Earth. The U. S. was still working on the civilian scientific Vanguard satellite of 3.25 pounds.

There was instantaneous reaction by the Senate Armed Services Committee. Senator Lyndon B. Johnson, Chairman of the Preparedness Investigating Subcommittee, brought forceful leadership to this challenge. 1/ and on November 25, 1957 began the basic investigation of the nation's resources for achieving a superior space program. The "Inquiry into Satellite and Missile Programs" 2/ called upon experts representing government and industry, science and technology, education, military and civilian fields of knowledge. These hearings continued through November and December and into January 1958, recording 2,476 pages of the facts essential for understanding the total situation as a basis for planning the future. The hearings were conducted in an emergency atmosphere of deep concern with the status of U. S. national defense.

An understanding of the comprehensive nature of the task was revealed by the testimony of scientists and engineers. Instead of fearing rockets as weapons, they hailed them as advanced technology for producing many benefits for mankind. Thus, the problem that required solution was two-dimensional: to preserve outer space for peaceful exploration and uses, and prevent its becoming a new arena for warfare. This situation was a classic case of presenting a choice between good and evil.

It became apparent that organization of the resources required for a space program would need to take the following factors into account: (1) satellites are inextricably international as they orbit over national boundary lines in 90 minutes or less; (2) all spacecraft require communications to send and receive information from the earth; (3) a network of tracking stations is essential and requires international agreements; (4) the geostationary orbit used for global communications has been declared a "limited natural resource", already organized by the International Telecommunication Union: (5) nations must guard against pollution, contamination of the Earth and outer space, including celestial bodies; (6) provision must be made for protecting the health and safety of astronauts; (7) measures must be taken to cope with interference from space debris; (8) space vehicles can operate and produce benefits only when they comply with the specific rules of space technology and science; and (9) space exploration is expensive but large projects can promote patterns of international cooperation.

As the configuration of the total problem took shape during the Preparedness Hearings, major issues arose. What should be the roles of military and civilian organizations? How should national and international aspects be administered? How should the Congress be organized to handle space legislation?

On November 21, 1957 the Rocket and Research Panel submitted a proposal for creating a National Space Establishment, an independent civilian agency separated from the military and funded on a long-term basis. A similar proposal had been sent to President Eisenhower by the American Rocket Society on October 14, 1957, only ten days after the Sputnik launching. The two proposals were combined and on January 4, 1958 the Congress was urged to establish a new civilian space agency authorized to conduct manned and unmanned space missions, consider a permanent base on the Moon, flights to Mars and Venus, and develop a variety of peaceful uses. Scientists and engineers promised international leadership on this "endless frontier" and identified benefits which could be expected from space activities: 3/

> There will be a rich and continuing harvest of important practical applications as the work proceeds. Some of these can already be foreseen -- reliable short-term and long-term meteorological forecasts, with all the agricultural and commercial advantages that these imply; rapid, long-range radio communications of great capacity and reliability; aids to navigation and to long-range surveying; television relays; new medical and biological knowledge, and so forth. And these will be only the beginning. Many of these applications will be of military value; but their greater value will be to the civilian community at large.

The Subcommittee concluded that the satellite was not yet a weapon, the Soviet Union had led the world into outer space, and it was now essential for the United States to make a tremendous effort to organize the resources for achieving preeminence in space. A finding was that the "same forces, the same knowledge, and the same technology which are producing ballistic missiles can also produce instruments of peace and universal cooperation": communications, meteorology, navigation and remote sensing. The responsibilities of the Subcommittee were limited to defense and it was necessary to create a special committee to handle the civilian aspects of space legislation. 4/

Investigation by House and Senate Committees

On February 6, 1958, the Senate established the Special Committee on Space and Astronautics with Senator Lyndon B. Johnson as chairman. Membership on the Committee was determined by the concern of existing committees with the variety of subjects involved in space applications that could cut across their jurisdictions. To solve what could become a complicated legislative problem, action had to be taken to avoid dispersal of the legislative process within the Congress, fix responsibility, and give outer space matters the highest priority.

In the House of Representatives a similar "blue ribbon" Committee on Astronautics and Space Exploration was created on March 5, 1958 with the Majority Leader, Hon. John W. McCormack as Chairman.

While these committees were considering how to organize the Executive Branch, Congress moved swiftly to pass interim legislation to speed U. S. space development.

The Secretary of Defense was authorized to engage in advanced research projects leading to the creation of the Advanced Research Projects Agency (ARPA), and for one year to be responsible for space projects as designated by the President 5/ The Supplemental Military Construction Authorization Act provided $10 million in transfer authority to the Secretary of Defense for advanced research. 6/

Among the organizations in a unique position to furnish basic information and advice on planning for the future development of space activities were the National Advisory Committee on Aeronautics (NACA), the International Geophysical Year (IGY) coordinated by the International Council of Scientific Unions (ICSU), and the Department of State on U. S. foreign policy for arms control.

Congress established the National Advisory Committee on Aeronautics (NACA) on March 3, 1915 as an independent agency reporting to the President. NACA was given authority to "supervise and direct the scientific study of the problems of flight, with a view to their practical solution...and to direct and conduct research and experiments in aeronautics." 7/ Research included missiles and manned aircraft; rocket research was concerned with engines for space and ballistic missions. Although a small agency with limited funds, NACA was successful with its expert staff in relations with the Department of Defense and as a link between the government and the aviation industry. 8/

The International Geophysical Year (IGY) was established by the international scientific community from July 1, 1957 to December 31, 1958 for scientific research of the world's total environment: the Earth, oceans, atmosphere, and outer space. This 18-month "year" was a period of peak sunspot activity favorable for interdisciplinary research. Two such studies had been made in what were termed polar Years in 1882-1883 and 1932-1933. Since that time, advances in communications and aviation technology had produced new tools for indepth research and 67 nations took part in the IGY which was organized by scientific disciplines, national committees, governmental sponsorship, and international coordination. 9/

In February 1953, the U. S. National Committee for the IGY was created by the National Academy of Sciences-National Research Council, the scientific U. S. organization that adheres to the International Council of Scientific Unions. The role of the government in administering U. S. contributions to the IGY was undertaken by the National Science Foundation which received an appropriation from Congress of $43,500,000 for IGY research. Additional financial and logistical support came from other government agencies involved with geophysical projects, e. g., the Department of Defense, Weather Bureau, Bureau of Standards, and Coast and Geodetic Survey. The linkage between the scientific community and the government was effectively managed, both nationally and internationally, by organizations with clear policy objectives, implementing programs, and adequate funding. 10/

The United States and the Soviet Union each made a commitment to launch a satellite during the IGY. The U. S. decision was announced by the White House on July 29, 1955: 11/

> ... the President has approved plans by this country for going ahead with the launching of small, unmanned earth-circling satellites as part of the United States participation in the International Geophysical

Year... This program will, for the first time in history, enable scientists throughout the world to make sustained observations in the regions beyond the earth's atmosphere.

Three days later, on August 2, 1955, the Chairman of the USSR Academy of Sciences, L. I. Sedov, stated, during a press conference at the International Congress of Astronautics, that the Soviet Union planned to launch a satellite. This announcement was confirmed by the Soviet National Committee for the IGY which described the scientific mission as measurement of atmospheric pressure and temperature, cosmic rays, micro-meteorites, the geomagnetic field and solar radiation. 12/

U. S. Arms Control Policy in the United Nations

U. S. policy on preventing rockets from being used as weapons in outer space, and limiting the new environment to peaceful uses, was being pursued in the United Nations and in diplomatic channels with the Soviet Union. In a memorandum to the United Nations on January 12, 1957, the U. S. called attention to the development of rockets and that it was "clear that if this advance into the unknown was to be a blessing rather than a curse, the efforts of all nations in the field need to be brought within the purview of a reliable armaments control system." On July 22, 1957, Secretary of State Dulles emphasized the opportunity to formulate in advance a system for ensuring beneficial scientific uses rather than destructive weapons. 13/

On November 14, 1957, the UN General Assembly passed Resolution 1148 (XII) calling for study of "an inspection system designed to ensure that sending objects through outer space shall be exclusively for peaceful and beneficial purposes." On January 12, 1958, President Eisenhower wrote Soviet Premier Bulganin proposing "that we agree that outer space be used for peaceful purposes ... Let us this time, and in time, make the right choice, the peaceful choice. 14/

By the time the Congressional committees began deliberations on future U. S. organization for space activities, the high priority objective had become peaceful purposes for the benefit of all mankind.

National Aeronautics and Space Act of 1958

While the House and Senate committees had been investigating the impact of Sputnik on the United States and the future for space development, analytical studies were under way in the Executive Branch. The President's Special Assistant for Science and Technology and the Science Advisory Committee combined their assessments with those of the Department of Defense, Department of State, the Bureau of the Budget and the National Advisory Committee on Aeronautics as a basis for recommendations on legislative proposals. 15/ On April 2, 1958, President Eisenhower sent his message to the Congress with draft legislation 16/ which was referred to the House Select Committee on Astronautics and Space Exploration and the Senate Special Committee on Space and Astronautics. There was agreement by the committees with three main issues: that a new civilian space agency be created; that NACA should be the nucleus for transfer to the new agency; and that the Department of Defense should have jurisdiction over projects primarily associated with military requirements. However, as legislative inquiries progressed, it became apparent that there were differences between the Congress and the Executive Branch on some of the major problems of organization that required solution.

The draft proposal assumed that all agencies would cooperate without overall coordination; NASA was to have an internal advisory board which met infrequently and had no authority over other agencies with space and space-related programs and budgets. If a question arose as to whether an activity was military or civilian, it was proposed that NASA "may act in cooperation with, or on behalf of, the Department of Defense." Although the draft preamble provided that NASA should cooperate with nations and groups of nations", there was no provision implementing this objective in the draft bill. 17/

The House Committee favored solution of the problem of overall coordination by civilian/military liaison committees of cooperation at the staff level with disputes settled by the President. The House Committee also recommended an internal Aeronautics and Space Advisory Committee of 17 distinguished members to advise the President. The Senate Committee, however, realized that the total U. S. space program involved coordination of a number of agencies and recommended a National Aeronautics and Space Board. This was changed by the Conference Committee to the National Aeronautics and Space Council whose function was to advise the President on the following duties: to survey all significant aeronautical and space activities, develop a comprehensive program, allocate responsibility, ensure effective cooperation and resolve differences. The Council was provided with a small staff. 18/ The objective of Congress was to ensure that all U. S. space activities would be handled at the highest Presidential level. The bill as finally enacted required the President to send Congress an annual report.

Both House and Senate Committees emphasized the necessity for international space cooperation. On January 14, 1958, Senator Johnson called for world leadership by the United States in the new dimension offered by space exploration: 19/

> We should, certainly, make provisions for inviting together the scientists of other nations to work in concert on projects to extend the frontiers of man and to find solutions to the troubles of this earth... it would be appropriate and fitting for our Nation to demonstrate its initiative before the United Nations by inviting all member nations to join in this adventure into outer space together.

On May 13, 1958, Congressman McCormack introduced a House Concurrent Resolution calling for the Peaceful Exploration of Outer Space and expressing the sense of the Congress -- 20/

> That the United States should seek through the United Nations or such means as may be most appropriate an international agreement providing for joint exploration of outer space and establishing a method by which disputes which arise in the future in relation to outer space will be solved by legal, peaceful methods, rather than by resort to violence.

The Concurrent Resolution was passed by the House on June 2, 1958 and by the Senate on July 23, 1958.

Emphasis on international space cooperation was further provided in the NASA Act by the Declaration of Policy and Purpose:

> The Congress hereby declares that it is the policy of the United States that activities in space should be devoted to peaceful purposes for the benefit of all mankind. (Sec. 102 (a)).

Furthermore, the aeronautical and space activities shall be conducted so as to contribute materially to --

> Cooperation by the United States with other nations and groups of nations in work done pursuant to this Act and in the peaceful application of the results thereof. (Sec. 2 (c) (7)).

Section 205 of the Act provides for international cooperation:

> The Administration, under the foreign policy guidance of the President, may engage in a program of international cooperation in work done pursuant to this Act, and in the peaceful application of the results thereof, pursuant to agreements made by the President with the advice and consent of the Senate.

When President Eisenhower signed the NASA Act on July 29, 1958, he stated that 21/

> I regard this section merely as recognizing that international treaties may be made in this field, and as not precluding, in appropriate cases, less formal arrangements for cooperation. To construe the section otherwise would raise substantial constitutional questions.

When the Senate Committee issued its Final Report, it recommended that the

> Congress should be kept informed of progress being made in studies undertaken by the United Nations Ad hoc Committee on the Peaceful Uses of Outer Space... Particular attention should be paid to preserving and extending the patterns of cooperation which were formed during the International Geophysical Year... The special committee commends the National Aeronautics and Space Administration for establishing an Office of International Cooperation... and recognizes the need of the Administration to provide for various types of cooperation as approved by the President. 22/

Thus the way was open for NASA to engage in many kinds of scientific bilateral and multilateral projects with other nations, leading to hundreds of projects with over 100 nations and international organizations.

Senator Johnson was determined that outer space activities receive continuing priority at all governmental decision levels. On August 21, 1958, two appropriation bills, one on military construction and the other on supplemental appropriations, were amended to provide for an annual authorization of funds for NASA. The requirement was at first temporary but became a permanent feature of the legislative process for NASA's activities: 23/

Notwithstanding the provision of any other law, no appropriation may be made to the National Aeronautics and Space Administration unless previously authorized by legislation hereafter enacted by the Congress.

Congressional Organization 1958 24/

The work of the Senate and House Special and Select Committes ended with the passage of the NASA Act and it was necessary to consider how Congress should be organized for jurisdiction over future space legislation. Four options were examined: (1) to create a new joint committee; (2) to divide jurisdiction among existing committees; (3) to refer space legislation to the Joint Committee on Atomic Energy; and (4) to create new standing committees in the House and Senate. The fourth option was chosen and the House Committee on Science and Astronautics was established on July 21, 1958. The Senate created the Committee on Aeronautical and Space Sciences on July 24, 1958. This action was unusual and underscored the intent of Congress that all space activities must receive specific unified attention according to the bicameral system.

The United States and United Nations

In little more than a month after creating NASA with the strong statement of U. S. policy in favor of international cooperation, President Eisenhower asked the United Nations General Assembly to include on its agenda a draft U. S. resolution calling for the establishment of an Ad hoc Committee on the Peaceful Uses of Outer Space. This Committee was to report on United Nations' resources and organizations which could advance cooperation among nations to avoid national rivalries and preserve outer space for peaceful uses; including identification of legal problems that might arise. The resolution advocated continuation on a permanent basis of the scientific research being carried on by the International Geophysical Year.

President Eisenhower invited Senator Lyndon B. Johnson to address the United Nations and lend his influence for adoption of the U.S. resolution. This occasion, when the President sent a plane to Texas to fly Senator Johnson to the United Nations, is a dramatic event in American history, demonstrating the unity of the Government when the Republican President joined with the Democratic leader of the Senate for international action on a critical U. S. foreign policy.

On November 17, 1958, Senator Johnson urged the United Nations to adopt the U. S. proposal for the Ad hoc Committee: 24A/

> ... if nations proceed unilaterally, then their penetrations into space become only extensions of their national policies on Earth. What their policies on Earth inspire -- whether trust or fear -- so their accomplishments in outer space will inspire also... Today outer space is free. It is unscarred by conflict. No nation holds a concession there. It must remain this way... We know the gains of cooperation. We know the losses of failure to cooperate. If we fail now to apply the lessons we have learned

or even if we delay their application, we know that the advances into space may only mean adding a new dimension to warfare. If, however, we proceed along the orderly course of full cooperation, we shall by the very fact of cooperation make the most substantial contribution yet made toward perfecting peace....

Nineteen other nations joined the United States in sponsoring the resolution 1348 (XIII) which passed the General Assembly on December 13, 1958. Membership on the 18-nation Ad hoc Committee was chosen on the basis of those most advanced in space technology and other nations representing fair geographical distribution. The U.S.S.R. (joined by Czechoslovakia, Poland, India and the United Arab Republic) would not participate because of opposition to majority voting. This hurdle was overcome, however, when agreement was reached on making decisions by consensus.

On December 12, 1959, the General Assembly adopted Resolution 1472 (XIV) creating the permanent United Nations Committee on the Peaceful Uses of Outer Space (COPUOS) whose membership has now grown to 61 nations. COPUOS established two subcommittees: the Scientific and Technical Subcommittee analyzes and reports on technical subjects which are then taken up by the Legal Subcommittee. This is the organization and process by which five space treaties have been formulated. The Conference on Security and Cooperation in Europe made a useful definition of the consensus process: 25/

Consensus shall be understood to mean the absence of any objection expressed by a representative and submitted by him as constituting an obstacle to the taking of the decision in question.

After attending many sessions of COPUOS and the Legal Subcommittee, this author observed: 26/

It is evident that consensus is a highly desirable way of achieving international accord because (1) the process of seeking agreement continues with patience and is not cut off suddenly by a vote which may defeat what might have come to fruition had more time been taken with the give and take process of consensus; (2) the situation may be such that a majority vote could not result in the adoption of a course of action, particularly if implementation of the decision in terms of funding, personnel and technological expertise, depended upon nations which had voted against the measure; and (3) group solidarity in decisionmaking ensures maximum compliance in establishing and maintaining an activity of general benefit. There is also a positive psychological effect when members of a group feel together with sympathy for differing viewpoints, motivated by a desire to bring about harmony in their collective judgment. If a member has not objected, a proposal can be adopted but this unspoken consent should not be interpreted as negativism; there is a positive willingness to settle the issue in question.

The creation of COPUOS as a permanent space organization within the United Nations advanced the movement toward international space cooperation for protecting outer space from conflicts. The method of establishing the facts of operational space technology as a basis for formulating legal principles, and the process by which consensus was obtained on decisions by political representatives strengthened compliance with essential guidelines. The United Nations became the forum for coalescing action at a critical time in history. Strong agreement on the objective of maintaining outer space for beneficial uses kept nations negotiating with patience until their disagreements were overcome. Within this UN context which the United States worked with dedicated leadership to establish, the UN COPUOS formulated the basic Treaty on Principles Governing the Activities of States in the Exploration and Use of Outer Space, including the Moon and Other Celestial Bodies, 27/ now ratified by 92 nations which comply with its science/technology oriented guidelines. From the provisions of this treaty, and to keep abreast of changing conditions of space science and technology, four more treaties have been spun off: the Agreement on the Rescue of Astronauts, the Return of Astronauts and the Return of Objects Launched into Outer Space, 28/ Convention on International Liability for Damage Caused by Space Objects, 29/ Convention on Registration of Objects Launched into Outer Space, 30/ and the Agreement Governing the Activities of States on the Moon and Other Celestial Bodies. 31/

In addition, Principles were adopted on four subjects: The Declaration of Legal Principles Governing the Activities of States in the Exploration and Use of Outer Space, with outstanding U. S. leadership, was adopted in 1963; the Principles Governing the Use by States of Artificial Earth Satellites for International Direct Television Broadcasting in 1982; Principles Relating to Remote Sensing of the Earth from Space in 1986; and Principles Relevant to the Use of Nuclear Power Sources in Outer Space in 1992.

All the UN-formulated space treaties have gained acceptance and compliance except the Moon Agreement which was opened for signature on December 18, 1979 but in 18 years has been ratified, and is in force only, for nine States: Australia, Austria, Chile, Mexico, Morocco, The Netherlands, Pakistan, Philippines and Uruguay, No space-faring nation has ratified this Agreement, largely because of its provision that the Moon is the "common heritage of mankind" linked to criteria to establish an international regime when its natural resources are about to become exploitable. It is evident that the perception of the problem and its matching solutions, needs to be updated in accordance with advances made in space science and technology as well as changing economic and political factors.

During the past 40 years we have been successful in extending international law into outer space, and to such an extent that we have created a special branch of space law. We do not need a police force to ensure compliance with these ratified legal provisions because they have been shaped to conform with the operational imperatives of space technology. And it should be noted that U. S. ratified space treaties have become the supreme law of the land according to the U. S. Constitution.

On December 31, 1958 the Senate Special Committee on Space and Astronautics published "Space Law: A Symposium" which was an expression of continuing concern with legal guidelines for orderly conditions in outer space. 32/ From time to time this publication is updated by the Senate Committee responsible for authorizing space activities.

Conclusions

If the Vanguard had been orbited before the Sputnik, we would not have been able to establish a coordinated national and international framework for developing the beneficial uses of outer space as quickly and successfully as we did; instead, we might have drifted into a period of instability without decisiveness on creating the conditions essential for maintaining outer space for peaceful uses. Vanguard first would have been an historical event hailed by engineers and scientists as tools for their projects, and it would have made front page news. But it would not have electrified the world; the U. S. public, in particular, would have taken for granted that we would be first in rocketry; and most of all, it would not have aroused the fear of orbiting Soviet weapons that galvanized political decisionmakers to take immediate action for achieving U. S. preeminence in outer space. Had the second Sputnik of 1,120 pounds been orbited after Vanguard, it would have aroused fear of those responsible for U. S. national defense, but information falling on the public in pieces over a period of time would not have had the same psychological impact on all groups simultaneously as happened with the timing of Sputnik.

The Sputnik was launched at exactly the right time, when space rocketry was in its earliest stages, to alert nations to the necessity for containing this global mechanism within the bounds of civilization. By striking such a complete lightning blow, the Sputnik had a unifying effect, drawing together all the societal elements needed for finding ways to establish conditions to ensure peace and prevent war in this pristine environment. Groups that were working separately on space missions, national defense, arms control, and within national and international organizations, immediately fused on one objective: to maintain outer space as a dependable, orderly place for beneficial pursuits. They were strengthened by the nature of space science and technology which produces benefits only when operating under normal conditions. No one wanted any disruption of space communications, which became a multi-billion dollar industry, weather predictions which saved lives and property, disaster relief, and the many practical uses of remote sensing.

The quickened pace of political action in 1957-1958 launched a regime that preempted outer space for the peaceful exploration and uses that we have enjoyed for 40 years.

In 1957 the United States had all the resources needed for developing a superior space program.

First, there was a thriving aviation industry ready to expand into aerospace.

Second, the National Advisory Committee on Aeronautics (NACA), a government agency already working on problems of flight and outer space, could be expanded into a new civilian space agency. NACA already had excellent relations with the Department of Defense and the aviation industry, and there was agreement that the Department of Defense could not develop all the civilian applications but needed jurisdiction over military space matters.

Third, the International Geophysical Year was organized by scientists and engineers working on outer space projects, and with a strong U. S. contingent.

Fourth, political leadership rocketed immediately on the day after Sputnik was launched. Senator Lyndon B. Johnson began planning for the full-scale investigation "Inquiry into Satellite and Missile Programs" which began in November 1957. He had the full support of the Chairman of the Senate Armed Services Committee, Senator Richard B. Russell. Senator Johnson was a driving force for speeding U. S. space objectives. On January 31, 1958, the U. S. orbited Explorer I, the satellite whose data led to the significant discovery of the Van Allen radiation belts.

Fifth, there was harmony between the Executive and Legislative Branches of the government; between the Republican President and the Democratic Congress on national and international space objectives.

Sixth, to implement space policies, organizations were quickly formed in the Executive Branch by dividing civilian and military uses and providing overall coordination; in the Congress with new standing committees; and in the United Nations by establishing the Committee on the Peaceful Uses of Outer Space. National and international organizations expanded to use space technology to improve functions they were already performing, i. e., the International Telecommunication Union (ITU), the World Meteorological Organization (WMO), and their counterparts in national governments.

During four decades of space development, some institutions made changes in their organization charts, but the general framework for achieving overall goals continued. Such changes are not unique to outer space and often occur because different managers like to alternate between centralization and decentralization practices.

As long as rules and regulations are effectively coordinated with scientific and technical facts, we can expect to maintain peaceful non-violent conditions. Difficulties could develop, however, if special groups were allowed to promote political and economic philosophies which ignore physical facts.

We have succeeded in expanding international law into outer space so that a new branch of international space law has been formed. Space law is remarkably self-enforcing because engineers and scientists have contributed to its formation by specifying the conditions they must have in order to operate and communicate between the Earth and outer space.

NOTES

1. Wilson, Glen P. Lyndon Johnson and the Legislative Origins of NASA. Prologue Quarterly of the National Archives. Winter 1993, Vol. 25, No. 4. 363-372.

2. Inquiry into Satellite and Missile Programs. Hearings before the Preparedness Investigating Subcommittee of the Senate Armed Services Committee. 85th Congress, First and Second Sessions. Part I, November, December 1957; Part 2, January 1958; Part 3, February, April, July, 1958. 2,476 p.

3. Compilation of Materials on Space and Astronautics. Committee print No. 1. Senate Committee on Space and Astronautics. 85th Congress, 2nd session. March 27, 1958: 14-19.

4. Inquiry into Satellite and Missile Programs, Part 3, op. cit., 2428-2429.

5. Public Law 85-325 (H. R. 9739), February 12, 1958.

6. Public Law 85-322 (H. R. 10146), February 11, 1958.

7. 50 U.S.C. 151.

8. NACA Research into Space. House Select Committee on Astronautics and Space Exploration. Hearings on H. R. 11881, 85th Congress, 2nd Session. April 1958: 404-410. See also, Resume of NACA Space Research Contributing to the United States International Geophysical Year Program: 457-458.

9. Historical Background and International Organization of the International Geophysical Year. Hearings before the Senate Special Committee on Space and Astronautics on S. 3609, National Aeronautics and Space Act. 85th Congress, 2nd Session, Part 1. May 1958: 132-136.

10. Sullivan, Walter, Assault on the Unknown: the International Geophysical Year. New York, McGraw-Hill, 1961: 27. See also, Annals of the International Geophysical Year. Vol. II. London, New York, Pergamon Press, 1959: 258-260.

11. Department of State Bulletin, August 8, 1955: 218.

12. The U. S. Rocket-Satellite Program for the IGY, a talk given on Nov. 9, 1955 by H. E. Newell, Jr. Third meeting of the Comite Special de l'Annee Geophysique Internationale (CSAGI) Brussels. Annals of the International Geophysical Year. London, Pergamon Press, 1959. Vol. IIA,: 267-271. Also, CSAGI Conference on Rocket and Satellite Observations. Report by Dr. Joseph Kaplan, Chairman of the U. S. National Committee for the IGY. September 11, 1956. Ibid., 300-310.

13. Documents on Disarmament, 1945-1949. Department of State Publication 7009, 1960. Vol. II,: 733, 832.

14. Ibid., 914-915; 938-940.

15. Killian, James R., Jr., Sputnik, Scientists and Eisenhower. 1977, MIT Press, Cambridge, Massachusetts. 315 p.

16. Hearings before the House Select Committee on Astronautics and Space Exploration, 85th Congress, Second Session. on H. R. 11881. April-May 1958 1542 p. See also, The Problems of Congress in Formulating Outer Space Legislation on pp. 5-10.

Hearings before the Senate Special Committee on Space and Astronautics, 85th Congress, Second Session on S. 3609. Parts 1 and 2, May 1958. 412 p.

17. Final Report of the Senate Special Committee on Space and Astronautics. Pursuant to S. Res. 256 of the 85th Congress. 86th Congress, First Session, Senate Report No. 100. March 11, 1959. 76 p.

18. Conference Report on the National Aeronautics and Space Act of 1958. House of Representatives, 85th Congress, Second Session, Report No. 2166. July 15, 1958. 25 p. All the changes are listed in this report.

19. Address by Senator Johnson before a meeting of the Columbia Broadcasting System Affiliates, Shoreham Hotel, Washington, D. C. January 14, 1958.

20. Final Report, Op cit., p. 17.

21. White House Press Release, July 29, 1958.

22. Final Report, Ibid., pp. 28-31.

23. Public Law 86-45, Section 4, June 15, 1959 (73 Stat. 75, 422-460. See also, National Aeronautics and Space Act of 1958, as amended, and Related Legislation; Senate Committee on Commerce, Science, and Transportation, 95th Congress 2nd session. Committee print. December 1978. 185 p. See also Eilene Galloway: The United States Congress and Space Law in Annals of the Air and Space Law Institute, McGill University, Vol. III, Montreal, Canada, 1978: 395-407.

24. Final Report, Op. Cit., Congressional Committee Organization and Jurisdiction, pp. 21-48.

24A Final Report, pp. 58-62.

25. A history of the Ad hoc Committee on the Peaceful Uses of Outer Space and events leading to the creation of the permanent Committee will be found in International Cooperation and Organization for Outer Space. Staff report for the Senate Committee on Aeronautical and Space Sciences. Senate Doc. No. 46, 89th Congress, First Session; 183-193 (1965). See also Conference on Security and Cooperation in Europe, Final Act 6. Rules of Procedure (69) 4, August 1, 1975.

26. Galloway, Eilene. Consensus Decisionmaking by the United Nations Committee on the Peaceful Uses of Outer Space. Journal of Space Law, University of Mississippi, Vol. 7, Spring 1979, No. 1. pp. 3-13.

27. This Treaty entered into force on October 10, 1967. 18 UST 2410. TIAS 6347, 610 UNTS 205.

28. 19 UST 7570; TIAS 6599; 672 UNTS 119.

29. 24 UST 2389; TIAS 7762; 961 UNTS 187.

30. 28 UST 695; TIAS 8480; 1023 UNTS 15.

31. 18 ILM 1434; 1363 UNTS 3.

32. Space Law: A symposium. Senate Committee on Space and Astronautics. Committee print. 85th Congress, Second Session. December 31, 1958. 573 p.

****

Dr. Eilene Galloway retired from the Congressional Research Service as Senior Specialist in International Relations (National Security). Following the launching of Sputnik, she was appointed Special Consultant by Senator Lyndon B. Johnson in 1957-1958 when he was Chairman of the three Senate committees that held hearings on the missile-satellite situation and organization of the National Aeronautics and Space Administration. She continued as Special Consultant to the Senate Committee on Aeronautical and Space Sciences. She was also Consultant to Hon. John W. McCormack, Chairman of the House of Representatives Committee on Astronautics and Space Exploration during the hearings on the NASA Act of 1958. She has been U. S. Delegate to the United Nations Committee on the Peaceful Uses of Outer Space and its Legal Subcommittee, and as Official Observer for the International Astronautical Federation. She received the NASA Public Service Gold Medal Award in 1984. She is Honorary Director of the International Institute of Space Law and Trustee Emeritus of the International Academy of Astronautics. She is a Fellow of the American Institute of Aeronautics and Astronautics and the American Astronautical Society. She is at present a member of the NASA Advisory Committee on the International Space Station.

Her biography is in Who's Who in the World, Who's Who in America, Who's Who of American Women, Who's Who in Science and Engineering, and Who's Who in the East.

Chairman ROHRABACHER. Thank you very much, and if anyone was just expecting a nostalgic look backwards, we got a—that was very good testimony and very provocative, and I'm sure we're going to have some serious discussion about that during the question and answer period, because I know that our next witness, Rick, probably has some questions about that, and some back and forth we can do as well.

Rick's a friend and an activist. Rick Tumlinson is a visionary, and the fact that he is such an aggressive activist sometimes makes people uncomfortable, but the fact is that he's an entrepreneur as well, and he leads the Space Frontier Foundation and he manages the Foundation for International Non-Governmental Development of Space which sponsors innovative space research, including asteroid surveys and a cheap access to space prize for very low-cost rockets. Rick, you may proceed.

TESTIMONY OF RICK NORMAN TUMLINSON, PRESIDENT, SPACE FRONTIER FOUNDATION

Mr. TUMLINSON. Thank you. I actually like what I'm hearing so far. The times they are a changing a bit. Now, if we can just make the reality meet the words.

This fall, we're going to be presented with two events that sort of epitomize the state of our space program some 40 years after its founding. Basically, 36 years after NASA achieved the ability to orbit John Glenn with great fanfare and trumpeting, this year, NASA is going to orbit John Glenn.

At the same time, very quietly, it will be divulged here and there, or leaked here and there, that the Space Station is once again over budget and going to slip. Thus, the dreams of Americans to go into space themselves will be, once again, forfeited.

These two events, contrasting as they do, demonstrate what is wrong with our space program today. We are basically, unfortunately, caught in a situation where we are going in circles. We spend our time aggrandizing the great moments of the past and putting off any moves into the future.

And why is that? I don't blame the Agency. I blame the fact that after an initial, very clearly laid out vision from which the Agency was started that led us to the moon, the vision stopped, and the Agency hasn't been given very strong, very clear directions ever since that time. We have basically taken the greatest team of explorers ever assembled in human history and told them that their job is to build buildings and operate a trucking fleet. This is an inappropriate role for an exploration agency, and, as some of the others have spoken about, NASA is not designed to do routine operations. In fact, what has happened is, the Agency, through having to deal with this set of circumstances, has become, at times, an inhibitor to the exploration and development of space.

So, what can we do about that? How can we turn that around? Because, frankly, I believe we can have all of the things that Mr. Goldin said. I believe that we can have human exploration of space. I believe we can have cities in space, cities on the moon. I believe Americans can walk, and that people from the rest of the world, can walk on the surface of Mars. I believe we can have reusable space transportation. I believe we can have clean, space solar power.

But we have to change what we're doing, and the way we have to change it is to reexamine the core assumptions in our space program as to who does what. Basically, we look out the door of this building and see how the American culture operates. And the way our culture operates is in a fairly clearly defined relationship between the private and public sectors.

This is what we have to do in space, and this is what I call the "space frontier agenda." We have to understand at the very core those two very different roles for the private sector and the public sector. And when it comes to space, I can summarize them very clearly: NASA's job is to explore; the people's job is to settle. NASA's job is to go beyond the hill, go beyond the horizon, tell us what they see there, bring back the information, and then let the American people go out into that domain and turn it into the extension of our civilization, the next place where people will go, live, and operate.

Once we decide that, once it becomes very, very clear that we have this division of labor, then we can begin to understand why we're in the position we are. We have taken this Agency and given it the wrong jobs. Then, when it was unable to perform those jobs, we condemned it for its inability to perform those jobs and told it, it can't go and do the right jobs. Once we get these definitions clear, we can move ahead, and we can move ahead strongly in a very exciting partnership.

First of all, I would lay out a few agenda items. Number one, I believe that it's the government's role to support technology development. The development of X vehicles and demonstration projects, such as those that have been proposed for solar power from space, are very exciting and should be continued.

We should privatize the Space Station as soon as we can and begin working on that right away. It is not NASA's job to build and operate a building.

We should privatize the Space Shuttle fleet as soon as possible. They've provided a great service by clearing the path on high technology. They should be allowed to provide a new service by showing us how we can begin to use and operate private space vehicles.

We don't need any more national spaceships, and we don't need a one-to-one follow on Reusable Launch Vehicles. What we need are vouchers. What we need are markets. What we need is competition. What we need is to free the new private launch firms that are out there, the reusable space transportation firms, to compete for government projects where national defense is not concerned.

We need to support these firms through regulatory relief and tax and investment incentives. We should not put in place such things as loan guarantees which have a tendency to favor certain companies over others, and, frankly, companies that have the money to be certifying the loan guarantees don't really need the money.

We need to privatize the Centers. The Centers are a great asset and could be institutionalized along the lines of the JPL Center in California, all around the country in private—university—government consortia.

Then—and this is the other part, the very interesting part and the exciting part—we need to change the laws so that the monies saved by these acts of privatization can be directly applied to exploration, the appropriate role for NASA.

I would like to see every dollar saved by privatizing the Station, Shuttle, and Centers poured into exploration. I believe that it is time for a President of vision, the next President, to stand up and say that we are going to put humans on Mars in 10 years; no long-term goals, let's make it exciting.

Let's put them there with money saved. We can do all of this without raising the NASA budget by one penny. Let NASA be the flagship for our Nation. Let NASA go do its job. Let NASA go to the far frontier, and let the near frontier, the area from the moon inward, become the domain that is led by the private sector with NASA acting as a good customer and supporting those activities wherever possible through purchases and other support.

I believe that we should have a return to the moon as soon as possible and that return to the moon should be a partnership. As the moon is on the edge of the near frontier, that domain I defined to be the realm of commercial activity, it is also a transition zone, and, therefore, on the moon we should encourage missions that involve transitional activities, such as a NASA-led exploration of the lunar surface that might be characterized through the use of data purchase, for example, from private firms or joint missions with private companies to the lunar surface. The discovery of ice on the moon has certainly given us the assets we need to move forward in that area. And we should encourage, through data purchase, private and NASA joint missions to near Earth objects such as asteroids and comets.

If we do all of these things, if we clearly define our roles, then we can have it all, and we can end the misunderstandings, we can end the bashing that goes on in both directions, and we can move out in a very strong way and give the American people the future they deserve and hope for in the 21st Century. Thank you.

[The prepared statement of Mr. Tumlinson follows:]

A Space Frontier Agenda

SPACE FRONTIER FOUNDATION

Testimony Before
The House Subcommittee on Space and Aeronautics
October 1, 1998

Rick N. Tumlinson

President - The Space Frontier Foundation

Executive Director - The Foundation for the International Non-governmental Development of Space (FINDS)

An Endless Frontier

Imagine a future without limits… a future where each tomorrow is greeted by hope, a future where the opportunity for a better life for each generation is no longer a question, and where a person's dreams are limited only by their own willingness to reach for them.

Imagine a future when Earth is freed from the pollution of fossil fuels and the fear of nuclear nightmares, where energy sources from space feed humanity's insatiable demand for clean unlimited power, and our resources are no longer torn from the heart of the Earth, but harvested from asteroids and comets.

Imagine a future when brave explorers spread out into deep space from a planet surrounded by bustling spaceports. A time when vacationers embark to Lunar resorts and space liners make the run to Mars filled with would be settlers who are putting it all on the line and taking the first steps to the greening of the Red Planet. A time when dozens of glorious crystal cities float between the worlds, housing tens of thousands of new solar citizens. This is the future in space Americans dream of, a powerful vision of a new civilization, spanning the solar system and beginning to reach out into the galaxy.

Once purely science fiction, thanks to the last forty years of work and exploration by NASA, we know these are *not* pie-in-the-sky fantasies. Through decades of research, much of it carried out by the people of our space agency, we *know* that these dreams are achievable. In fact, most of them could be created with technology already in hand or under development. I'll say it again. Thanks to the foundation created by our space agency in the last 40 years, these visions are possible. No magic elixirs are needed to make them happen. Forty years into our national space program we now know that we *can* do these things. What's more, in a world desperate for vision, hope and a better tomorrow for all its people, we cannot afford *not* to do them.

Yet, forty years into our space program you and I and our children have little more chance of being able to go into space and participate in creating this dream than we had at its beginning. After a glorious start, our people's opening of the frontier has now fallen victim to its own heroes, and a lack of a clear vision of the future on the part of our leaders. The space program that made these dreams possible can't deliver.

Going in Circles

This fall, America will be presented with two events that epitomize the state of our national space program. First, NASA will with great fanfare and self-congratulation launch an aging icon of its past glory to repeat his last performance orbiting the Earth. Then, with an embarrassed whisper, NASA will divulge yet another billion-dollar cost over run and another long delay in the construction of its alleged future, the International Space Station (ISS).

This counterpoint of events clearly illustrates the bankruptcy of our national space program, demonstrating its focus on reliving past glories as a substitute for real achievement. Once the greatest team of explorers in human history, today our space agency's actions belie its slogan of faster, better, cheaper, by spending a fortune to move as slowly as possible in an endless circular orbit.

To solve this problem, we must throw out our old model of space as a mere conduit for communications signals and occasional scientific playground for a few elite government employees and their chosen friends. In its stead, we should invoke our own history. We should look to the hard wrought partnership between the public and private sectors that made our nation great and allowed our ancestors to turn this "New World" into the incredible place it is, the home of history's freest and most dynamic civilization.

All of the Above!

There is a solution. One that makes the old debates about space moot. A solution that makes the false dichotomies of this vs. that destination or this vs. that project dissolve. We ***can*** send NASA to Mars and beyond *and* have space stations *and* cities on the moon *and* tourism *and* clean space energy - all without raising our space budget by a single dollar. If we will simply apply the lessons of our society, wherein a clearly defined relationship between the government and free enterprise produces enormous wealth on a daily basis we can begin to open the space frontier *now*. Based on a common sense application of what works here on Earth to what we do in space, this is what I call the Space Frontier Agenda.

A New Division of Labor

The first task on the Space Frontier Agenda is to clear up a few basic misunderstandings regarding the appropriate jobs of the government and private sector. It is this mix of wrong jobs vs. wrong institutions vs. wrong expectations that has produced the current mess in space policy.

In America and free societies everywhere, the government's role is that of rule maker, supporter, and provider of infrastructure and realizing our collective vision and identity. For example, the government may open new roads to serve the economic needs of the people. It may explore the land and finance the surveying and acquisition of rights of way and it may regulate the speed and types of vehicles on the highways, but it doesn't build and operate the vehicles that use them and it certainly doesn't build the towns they serve.

Why should space be so different? Don't the same rules apply?

The Right People - Wrong Job

NASA is in effect an institutionalized descendant of the Lewis and Clark Expedition, a federally chartered exploration and research project. Like Lewis and Clark, their primary activity should be to explore, survey and expand our knowledge of the frontier just beyond the borders of civilization. Unfortunately, in many cases, that is exactly what we have told them not to do.

The first decade and a half of its existence were characterized by stunning achievements in human exploration and planetary science. However, for the last 25 years NASA's human activities have been wrapped around two central projects - operating the space shuttle fleet and building the International Space Station - that are antithetical to what it was designed for, while all else has been starved into mediocrity.

Our leaders have ignored the evidence of history and saddled the members of the best exploration and science organization in history with exactly the chores they have neither the competence, passion, nor the cultural inclination to do - producing low cost and routine transportation and developing real estate. Ironically, due to its total failure to do these *wrong* jobs, the agency now doesn't have the ability or the credibility to do the *right* job: expanding our knowledge of the solar system and beyond and exploring Mars as a prelude to settlement.

The Division of Space

It is time to get things straight in space. When it comes down to the job description for our new space enterprise, let's keep it simple:

NASA's Job is to explore.

The People's Job is to settle.

Near Frontier: Private Sector Leads

I believe that NASA's Lewis and Clark's have done their job here in the neighborhood of Earth. This area - the Near Frontier - is comprised of the area from the Earth to the Moon, and the surprisingly large number of comets and asteroids that either inhabit or pass regularly through our neighborhood. From low Earth orbit, where Senator Glenn first flew almost 36 years ago and where Americans and Russians have lived for two decades, to geostationary orbit and its halo of communications satellites, it is the next step outward for our species, the next zone for expanded human activity. This area is unique in all the Solar system, since the costs of accessing it are far lower than other areas, much time has been spent exploring its potential and it is already home to early economic activity. NASA has done an excellent job exploring, surveying, characterizing and demonstrating the potential of this little piece of the universe, now it's time for the rest of the nation to do its job. The Near Frontier should be handed over to universities and private firms to explore and develop for human use as soon as possible.

Here, the government should create and apply laws and regulations that make space a safe place to live and work, just as it does here on Earth. At the same time, Congress and the White House should work together to streamline regulations, clearing the way for innovation and swift development.

Space should be declared an enterprise zone, and the widest possible range of tax incentives and regulatory relief should be applied to encourage the investment community to provide the capital needed by those trying to build new space businesses. The government should also learn to become a good customer, sticking with agreements, and avoiding sweetheart deals or picking market winners. NASA must also abandon acquisition practices that are process oriented rather than based on deliverables, a major cause of cost over runs. It simply doesn't matter where or how a particular project is built, or how needed information is acquired, as long as it is delivered on time and at the price agreed to.

Far Frontier: Public Sector Leads

The Far Frontier is currently defined as the rest of the Universe, including Mars and beyond. Of course it is a flexible border, moving ever outward as the private sector follows the paths blazed by our explorers, learns way to create new wealth from what they have discovered and turns these areas into the new edge of the human domain.

Our activities in this realm can be broken into two types; pure scientific inquiry such as that now carried on by the Hubble Space Telescope, and exploration for both science and as a precursor to human settlement or use. This is how we should approach Mars and the rest of the Solar system, and makes a joke out of the old argument regarding humans versus robots as explorers. (Once human settlement is the established goal, then bottom line considerations will usually dictate early robotic exploration, followed by people ASAP.)

On the far frontier pure science should be funded through grants to institutions such as universities, with projects such as Hubble handed to a university or consortium to operate. NASA should also work to create systems that place data as directly as possible into the hands of the people who need it, avoiding bureaucratic or funding based bottlenecks. One system in the works that would support this goal is the recent agency proposal to allow researchers to acquire information directly from probes by extending the internet into space.

NASA and other agencies should always seek to support the private sector in space, for example buying data for both scientific and settlement oriented exploration from private firms, or providing grants, and in some cases supporting mixed private and public sector missions.

The Government Role in Technology

The government has an important role in supporting the development of advanced technologies. NASA should support research that is beyond the financial scope of the private sector, including X-vehicles and technology demonstrations.

The difference between X-vehicles and operational systems is that X-vehicles are intended totally for use as technology test beds, are never intended to be operational and have no business plans attached to them. For example, the X-33, X-34 and X-38 are already providing useful information for those working on low cost space transportation in the private sector, and promise many more breakthroughs before they even fly. In the past the temptation (and to some extent the tradition) has been to let test series like these morph into development projects, as is the case with the X-33 and NASA's desire for this project to lead to a space shuttle follow on. Since the Space Frontier agenda calls for competition and open market purchase of these services to force costs down, this will be unacceptable. In all cases, once X-vehicle's test flights are completed, it is up to the private sector to apply the technologies they demonstrate to commercially viable products.

Technology demonstrations, such as the construction of a viable space solar power system, or a working terrestrial simulation and testbed for a Lunar/Mars base are also useful activities for the government to support. Working with industry partners NASA can focus brainpower and attention beyond the scope of any one firm to push back the technological frontier as it does the physical.

The government should offer prizes for specific and clearly defined breakthroughs. Prizes can be used as magnets to draw genius out of our society. Prizes have a life of their own that crosses traditional lines and jumps over the walls of academic specialty. However, to be effective, such prizes should be focused, concise, and limited in scope.

The International Space Station (ISS)

We should take the first steps to privatize the International Space Station now. Government has repeatedly shown it is incapable of taking on construction projects here on Earth. Why should space be any different? From federal housing to downtown office buildings, to super colliders, government management of such fiascoes always results in embarrassing cost over runs and delays. This is exactly what has happened with the construction of the International Space Station, and should be no surprise to anyone familiar with government.

When seen in combination with the inefficiencies and high costs already manifest in space shuttle operations it should be clear to all that managing such a facility is not the right job for the agency. We should take this as a clear warning of what is to come if we continue down this path. We must solve this problem now rather than later, as it will only get worse and cost taxpayers more if we wait. What's more, the solution offered here has the potential to turn this slow rolling disaster into an economic gold mine.

The private sector should take over ISS as soon as possible. The new landlords might take the form of a public and private institute or consortium. They would be free to negotiate deals and contracts for services, set prices based on real costs, collect rents and sign binding leases with paying tenants, including national governments. (Of course the founding station partners would be given long term renewable leases at no cost.) I believe the private sector is up to the task, and if given a clear and unambiguous go ahead, would not only perform the task well, it would turn the station into a catalyst for whole new industries.

Government must support the recycling and development of space real estate. If we are to believe NASA's own hype that low Earth orbit is going to become the home to a wide variety of scientific and commercial activities, then we can expect the needs for viable space real estate to grow. In this light the current waste of such potential assets as the massive space shuttle external tanks that we routinely toss to burn up in space becomes obvious. Further, it should make us re-examine the extremely shortsighted decision to discard the Russian space station Mir.

According to almost every unbiased report, Mir is completely viable, dangerous to bring down and could for an amount of effort equal to or less than that required to de-orbit it be shut down and boosted into a high "sleeping" orbit. I can easily imagine some future private space flight company working with the Russians and a US/Japanese consortium to awaken the station, refurbish it and turn it into a private micro-gravity laboratory, hotel or resort.

Alpha Town

Here on Earth, we have learned that it is never one single product that sustains the opening of frontiers or the creation of viable towns and cities, but the interaction of all types of human activities. The same is true in space. In previous testimony before this committee I laid out a means for ISS to support the development of the first human town in space "Alpha Town".

The "Alpha Town" idea is based on the historic role federal frontier outposts have served as the catalysts for civilian communities by acting as anchor tenants and early market makers. For thousands of years, nations have placed outposts and forts just beyond their borders in strategic locations. These facilities also became anchors for new transportation routes. Purchasing their goods and services from the local hunters and trappers, the outpost soon attracts others wishing to sell their goods to the government. These people soon begin to trade with each other, and to use the established transportation routes to export their goods back to the heartland. As time passes, the outpost or fort becomes an economic center, then a town, and often, comes to form the heart of a modern city. (Where the old fort sits downtown as an historic monument, its job done.)

As an example, as ISS utilization increases, the former federal space station and other commercial space facilities such as a privatized Mir, and recycled external tanks will not only feed their own markets, but those providing services to them will become markets for each other. More activity will drive the need for more transportation, more services and more volume within which to grow - causing more competition among providers of these goods and services and lowering costs still further - which drives more growth, and so on...

If the Space Frontier Agenda is adopted, transportation costs should drop sufficiently that low Earth orbit (LEO) will become a community of diverse economic activities, from Earth studies to energy production, satellite repair to processing and science to entertainment, with the ISS as its hub. Our first town in space, "Alpha Town" will be born.

Cheap Access to Space

The development of cheap, reliable and regular transportation to and from space is THE key requirement for opening the space frontier. Once again, there are strong mutual interests between the private and public sectors to be satisfied, once again, there is a chance for a partnership, and once again there is the chance to create new industries and jobs. And unfortunately, once again we are faced with a government monopoly - this time operating the only human capable space transportation system in the United States.

Privatize the Shuttle Fleet

The era of national spaceships must end. The US government must never again own and operate a national fleet of vehicles for civilian purposes. Deprived of market forces or meaningful competition, the shuttles were the created in the old government contracting system. Also, as a result of endless compromises between NASA's perceived needs and that of the military, the space shuttle fleet lost its original pedigree as a low cost space transportation system before it ever left the drawing board. However, it is the only U.S. system for carrying people to and from space currently flying. As both a symbol and a test case for the transition to commercially driven and operated human space transportation, the shuttle fleet can pave the way.

It is time to turn over the shuttle fleet to private operators and allow them to carry anything or anyone they like. Although NASA has, with its usual great fanfare created a thin veneer of privatization in shuttle processing, most of the activities it ostensibly gave away have already been taken back. This is unacceptable. We need real top to bottom privatization of the fleet and we need it now. The new operators of ISS can deal directly with the company running the shuttle fleet to cut whatever deals make the most economic sense, and NASA will finally be free to buy its rides on the shuttle or any other vehicle, rather than fulfilling artificial flight rates. True, the shuttle's operators will eventually have to compete against the new generation of space transpiration companies, including the potential descendants of the X-33 and X-34, but they will have several years to prepare and carve out their own market niche.

The transition to private operations will allow the shuttle fleet to perform one last great service for America. The space shuttles broke technological ground by allowing us to take the first steps towards reusable spaceships. Now they can blaze the regulatory, financial and economic trail to space by carrying the first U.S. paying passengers into space, setting standards for citizen training and access, and pioneering the transition from federal space transport to civilian space lines.

Support new space transportation companies.

The federal government must implement regulatory and investment incentives for the new space transportation industry and allow it to compete fairly for all government payloads, including those of the Department of Defense. One of the most exciting developments of the last few years is that several new space transportation companies crossed the threshold from being business plans to businesses, with capital, staffs and engineering either soon to begin or already under way. The Foundation is working to cut a path for these new companies, as we believe they hold our greatest hope in the quest for cheap access to space.

Unencumbered by traditions, bureaucracies and structures designed to siphon tax dollars rather than realize profits, these firms are where innovations and new ideas can be born and tested on the anvil of the market system. But they face enormous challenges on the road to success. Ironically, many of which were created by our own government.

We must create a favorable regulatory environment if new space transportation firms are to survive. Let me give you a recent example of what these companies are up against. Although we all know that reusable rockets are one of the near term keys to the opening of space, there were laws on the books which allowed rocket builders to launch their rockets, but banned them from bringing them back down. Imagine, its like someone from the FAA running alongside Orville Wright at Kitty Hawk and telling him he could not land. In this case congressional wisdom saved the day and the law was changed, proof that leadership by Washington can have a positive effect. It is time to review, revamp or remove our laws and policies regarding space transportation. Government must work with this new industry as a partner, to ensure safety and also to encourage innovation.

We must offer tax and investment incentives to those investing in space transportation. As I discussed earlier, one tool of government is the ability to support new industries through the offering of tax and investment incentives. Just as it supports the development of new "land" and companies that provide services to space facilities, the federal government should create a package of tax and investment incentives to encourage investment in space transportation.

We must avoid the market warping effects of loan guarantees. Federally subsidized loan guarantees are being promoted by one or two firms who's planned vehicles will require the most extreme outlays of cash, and are the most financially dubious of all the ideas being proposed, when costs vs. risk vs. returns are considered. All of the rest of the new space transportation firms agree that this concept is very bad for this fledgling new industry.

Loan guarantees may sound good on their face, but they are usually the last resort of those whose business plans have been rejected by Wall Street. It can also be expected that any criteria for their disbursement will be established by a few inside players, and will warp the market in favor of those firms with the most assets and best connected lobbyists, not the best ideas. If government payloads are offered in a fair and consistent manner to the market, and tax laws made favorable to investors, the best companies will have no trouble raising capital.

Privatize the Centers

The government should privatize the NASA centers so they can act as true research centers and incubators for new space industries. The development of commercial and other private sector activities in space will drive the growth of supporting industries here on Earth, industries that will cluster near space facilities. There will be a great need for test, training, processing and other specialized capabilities. This is where another legacy of our first forty years in space can provide a major kick-start to the opening of the frontier.

Designed to spread out political support for the early space program, and sometimes the result of what must be called pork barrel politics, the vast space center-based infrastructure developed by NASA is filled with redundancies and inefficiencies. It is dominated by turf grabbing and politics and has bred a culture of division that inhibits progress. We are too often faced with the specter of this or that politically powerful space center actually competing with or killing private sector ideas as a means to retain their own power and funding. The centers should be spun off to partner with local industries and universities wherever possible, as the cores of future space industrial and research parks, with the successful CalTech/JPL model as a minimum standard.

Spend the Money on Exploration

We must take the money saved from privatizing operations in the Near Frontier and apply it to a greatly expanded program to explore the Far Frontier including Mars, the rest of the Solar System and beyond. This is one of the most controversial elements of the Space Frontier Agenda, but it is also one of the most important. It is absolutely key to the Space Frontier Agenda that we not simply end our federal program of supporting exploration and discovery. In fact, the goal is to greatly expand this most inspirational and educational of human activities **without spending more money.**

By moving inappropriate activities such as operating the ISS and space shuttle to the private sector and lowering overhead burdens imposed by the current centers, billions of dollars will be freed up for areas of research and exploration starving for support. The new NASA this re-organization creates will be leaner, more focused and ready to take on the challenges it was designed for.

The role of the new NASA must be to send its robotic and human explorers over the next hill, to support those who would bring us visions of new worlds and investigate new ideas beyond the pale of industry and corporate investments. The national program, once it is cleanly separated in role and expectations from the private sector, will act as a flagship, telling the world, investors and those who would become pioneers that we are moving ahead into this new millennium in a strong and confident manner. New discoveries in our solar system will pull humanity outward, and new discoveries beyond our home system, such as other living worlds orbiting other suns will certainly act as an impetus for our expansion outward into his new domain.

In the past, any savings the space agency was able to create through increases in efficiencies or through deals with private firms have vanished into the overall federal budget. This obviously does not encourage the type of innovation we need to realize the Space Frontier Agenda. Laws must be changed and enacted that allow the new NASA to transfer all money saved by handing off operational activities into exploration, technology development and science.

The Settlement of Mars

We must greatly expand and accelerate the exploration of Mars, particularly as it enables the settlement of Mars and the rest of the solar system. Money's saved from space station shuttle and center operations should be used to fund the development and demonstration of pioneering technologies that will enable the exploration and settlement of Mars. And yes, humans should go to Mars, as humans should go everywhere that it makes technical, economic, scientific, environmental sense to go. That's what an open frontier means.

In fact, if the rest of the Space Frontier Agenda is followed, in the very near future it will be time for a new president to announce that we are going to Mars. Although this idea has been tried before, it was within the old paradigm, and was based on a massive infrastructure build up that would have had our Lewis and Clarks building a massive support infrastructure on their way to the frontier. These needs will now be filled by the private sector. When this approach is combined with new low cost ideas for missions that are based on developing the means to live off the land, the opening of Mars becomes a viable idea.

The drive to open Mars to human settlement will fire the imaginations of our youth in a way that the more routine operational aspects of settling the Near Frontier will not. It is a symbol that will have a positive effect on all space activities, if it is part of the agenda I have outlined here. It will be seen as a national endorsement of space as a frontier, and it will be the most visible aspect of the government's role in the new space partnership I suggest.

In frank political terms, human exploration of Mars also provides the carrot needed to pull NASA's management, human space flight centers, the astronaut corps and its cheerleaders away from the Near Frontier.

I do not support a dramatic increase of NASA's budget that could be used to fund a massive "flags and footprints" Apollo-style program, as I believe it is politically unworkable, and does not lead to widespread, cheap, and permanent human settlement soon enough. Also, keep in mind that someday Mars too will be a part of the Near Frontier and NASA will have to move on yet again to the edge.

Return to the Moon

We must return to the moon, this time to stay. The moon was the first and primary focus of NASA and our space program. Although for the wrong reasons, it has now been rightly dropped from its central position on the agency's agenda. NASA and the federal government have done their job on the moon well, and it is time for them to move on. The next people stepping onto its surface will not be government employees. The moon lies on the outer edge of the Near Frontier, both physically and in the sense that it represents a transition zone from the market driven economy of LEO to the more publicly supported explorations of the Far Frontier.

Luckily, as NASA considers itself to have "done" the moon. Thus we are spared the usual battles over turf and fear of competition we see in the station, shuttle and other areas. We can use the moon as a test case for new partnering arrangements between NASA and industry, as well as the first place beyond Earth that we can apply the concepts of data purchase to support the first commercial space exploration firms.

There are already many ideas for commercial lunar activities, ranging from entertainment to astronomy; mining to commercially operated Mars Base training simulators. With the recent discoveries of large ice deposits concepts for the first Lunar colony are already being put together. Our government should do all it can to encourage and support private sector activities on the lunar surface. From purchasing data and acting as a partner with private firms and institutions on early transition projects to fighting for the maximum freedom of property ownership and resource development rights, the government still has an important role to play on the Moon.

While NASA is being re-organized and major shifts in priorities and plans are made, a very small amount of support for small proposed NASA/commercial Lunar missions could go a long way to laying the groundwork for the rapid development of the first human outpost on another world.

Near Earth Objects

We must support the purchase of data and the private development of NEOs. The threat from asteroids and comets has been the focus of the media this year, highlighting the need for a greatly expanded search for these objects, which could wipe out life as we know it. This search, with its broad societal implications, is the proper domain for the government to provide support. Although private groups such as FINDS and the Space Frontier Foundation's Watch project are trying to raise private funds for this important work, more help is needed, so we may find and understand what is out there.

I believe that one of the most important benefits of this search will be to expand our understanding of their promise to our civilization. Many of these objects are literally floating gold mines, continuing amounts of gold platinum and other precious metals that would stagger the imagination. They offer us the chance to greatly increase our wealth while moving environmentally destructive mining operations from the living Earth to these dead rocks.

As with the moon, the new NASA should support transitional missions, with large commercial participation in the form of partnerships or outright purchases of data. And just as in the case of the Moon, there are firms already well on their way to begin the exploration and exploitation of these exciting resources.

A Frontier Partnership

With the roles well defined and the goal for all clear we can at last embark on the opening of the frontier. The robots, lenses and heroes of the new NASA will peer and travel over the far horizon and tell us what they find there, and we the people will follow, creating new wealth and new worlds from what we discover.

Resurrecting Family and Community

The effect of the space frontier on American families and sense of community will be profound. Prosperity will increase, optimism will rise and new leaders appear in response the endless challenge provided by its opening. Our pioneering past will at last have a direct link to our future. The path blazed by Daniel Boone and Davy Crockett will continue onward and upward across the stars. The spirit of family and community will be resurrected as the frontier ethics of hard work and familial support are re-enforced through the basic need to survive and prosper while facing the harshest frontier ever conquered by humanity.

A New Millennium - A New Frontier

As we stand at the beginning of the millennium, it seems clear that we are a culture in transition. This century began with most of the world just beginning to break out of social, technological and economic systems that were unchanged for thousands of years. In a mere century we have totally reinvented our world. From horse powered transportation and communications based on word of mouth we have made it to the internet, space flight and a global community that has at one in the same time the ability to destroy itself or leap off the planet.

The opening of the frontier will mean a reversal of our seeming descent into the grey oblivion of limits. It will mean a reversal of all we know and understand about our relationship with each other, our world and our fellow life. Once the potential destroyers of our womb world we now stand at the heart of historical irony. We, the bringers of death to so many forms of life on this precious planet, may now have the capability to take life to worlds now dead.

...to take life to worlds now dead.....
There is in this act divinity. There is in us purpose.

From killers to creators...
From destroyers to cultivators.
This is the choice we face.

If we make the right one then someday trees will grow on the Moon, flowers will bloom on Mars, and fish will swim the seas of Europa.
And out there somewhere, on a sea of grass in a bubble of life, a child born in the asteroid belt will chase a butterfly that calls the space between worlds its home.... Looking at the pale blue dot in the night sky, that child will know the Earth is safe. Looking outwards at the rest of the universe, that child will know why we exist, the great promise our future holds, and smile.

Chairman ROHRABACHER. Thank you very much, Rick. At this point, I'd like to ask unanimous consent to insert into the record written statements that we have received from Keith Cowing, the Editor of NASA Watch; The American Institute of Aeronautics and Astronautics; and the National Space Society. Without objection, so ordered.

[The prepared statements of Mr. Cowling, the National Committee on Space, the National Space Society, and the United States Space Foundation follow:]

Written testimony submitted by
Keith Cowing, Editor, NASA Watch
before the
Subcommittee on Space and Aeronautics,
House Science Committees

"NASA at 40: What kind of space program does America need for the 21st Century?"

1 October 1998

Mr. Chairman:

Thank you for the opportunity to express my views on the future of NASA and our space program on the occasion of NASA's 40th birthday. I come to this task by a curious route - the Internet. I am a former NASA civil servant and space life scientist who now runs a small business that develops the content for Internet. I also run several news oriented websites at my own expense, one of which is called NASA Watch (http://www.reston.com/nasa/watch.html)

NASA Watch sprang out of an earlier, somewhat ad hoc Internet effort to provide information to NASA employees about looming threats of layoffs - something NASA was loathe to publicize. Since its debut several years ago, NASA Watch seems to have become a virtual destination where thousands of people in and around NASA visit daily to exchange (or just read) information about NASA. Much of the information on NASA Watch is composed of items NASA either does not bother to release or wishes would not be released.

I do not profess to be an expert in space policy and will readily admit that I usually just type what I think or feel as the thought occurs to me. Suffice it to say I used to work at NASA, and I talk to many people daily who still do. I am worried that our space program is not being operated or supported as well as it could be, and I decided to speak out.

Summary

My premise is simple: NASA excels when our nation has a clear, binding, national space policy. Absent such a policy, our space infrastructure fragments into dueling factions. Without a clear mission NASA now seeks to maintain its own existence with various missions serving as a means whereby to accomplish this end. Meanwhile, the relationship between all of the players has gotten worse over the years and is now highly dysfunctional.

This hearing is aimed at NASA and its future. Unless fundamental aspects of how NASA operates today are fixed, NASA will be unable to meet the challenges of the next century.

53-067 99 - 4

Indeed, one might question whether NASA in its current configuration is even the appropriate mechanism to manage our exploration of space.

I am not certain just what America's guiding vision for the exploration of space should be. All I know is that we are in desperate need of one.

My underlying assumption is that we as a nation, and as a species, are destined to expand outward from this planet to personally explore the universe. As such, we need an organization to guide the way. To date, this has been NASA. I see no reason why NASA should not continue to lead America's exploration of space. But in so doing, NASA needs to constantly adapt to changing pace and character of exploration. Below, I will deal with many of the problems I see as affecting NASA's current ability to perform its mission. These problems need to be addressed if we expect NASA to go further and learn more in space.

Visions and Frustrations

On the wall above the dais in the Committee's hearing room is a verse from Proverbs: "where there is no vision the people perish." NASA's core problem today is that it has no clear compelling vision to guide it. Nor has NASA had such a vision for decades. When NASA last had such a vision, humans went to the moon a mere 8 years after they decided to do so. We were guided by a clear and simple concept - one that sprang forth from the White House and was embraced and given form by the American people.

What we lack today, as we stumble ahead in our exploration of space, is such a clear goal and shared common vision. Our space activities have suffered greatly as a result. If a clear vision is not soon established for the NASA of the present - there may no longer be a need for NASA. Indeed we may find ourselves retreating from space instead of exploring it. The current way of doing things is broke and needs fixing.

Today, in the absence of a clear, unifying vision, everyone involved ends up creating and then defending their own vision if for no other reason that to justify their own continued existence. The net result is a constant clash of individual self-serving agendas, each held by one branch of government, agency, organization, or industry. Survival replaces progress and we all lose out.

Who is at fault for allowing this situation to happen? We all are.

NASA emerged in a time where ideas seemed to be simpler to express - and comprehend. Back then, we were in a race to the Moon against our clearly defined enemy (Russia), bound to a clearly defined deadline (1969), and in pursuit of a goal embodied within a dream, which had been shared by humanity for millennia. What could be simpler!

Politics, current events, and human yearnings all conspired rather effectively to get us to the moon - but this wonderful confluence of factors was unable to provide a compelling reason to stay there - or to move on. Indeed we began to disassemble this incredible capability within a few months of the first landing on the Moon. The vision that had propelled us to do things we thought impossible had evaporated in the bright light of our own success.

After Apollo achieved its primary goal, our space exploration infrastructure began to shift from one driven by a goal to one of self-perpetuity. Projects were now created to justify NASA's existence - not the other way around.

Mr. Chairman I am frustrated by all of this. Exceptionally frustrated. NASA and I grew up together. I was not yet three years old when NASA was born. Just as many people stop at midlife to assess where they've been and where they're heading, NASA finds itself doing much the same - only to be confronted with a midlife crisis. NASA is too preoccupied with justifying its own existence and putting out the brush fire du jour to take note of this crisis.

In its youth, NASA was an agency that commanded whatever resources it needed, leapt back into action after many a setback, and could be trusted at its word when it said that a thing could or could not be done. Now NASA is insecure and is subject to the whims of others, burdened with a host of operational responsibilities almost unknown in its youth. As such it is often focused inward upon its own survival with its original core research mission taking second priority. NASA needs to return to the things it does best and shed all other distractions.

I grew up in the 1960's expecting that we'd go to the moon to stay, and that we'd be on Mars by 1981 - just as Vice President Agnew and others had suggested as we began to land humans on the moon. As a teenager I saw the film "2001: A Space Odyssey" and thereafter firmly believed that I would reach middle age in a time when human civilization spanned two worlds.

Given the rate at which we progressed as we leapt from Alan Shepard's 15 minute ballistic flight atop a stepchild of the V-2 in 1961 to the first Saturn V launch in 1967, this was certainly within our grasp. This happened between the time I went from 1st to 7th grade. I guess I can't help holding a mental image of rapid and sustained progress in space propelled by a simple clear vision - for it was the very first impression of space I ever experienced.

Let me pause to note that what I have said thus far, and what I will say ahead, may sound na•ve and idealistic. In a town swarming with complicated ways of doing things, I'm sure it does. To be honest, Mr. Chairman, I am tired of listening to professional space pundits, politicians, and NASA Administrators (such as represented by some of the people assembled for this hearing) who try and explain why we have or haven't done things in space - each of whom is unwilling to admit that they are at fault for taking what could be one of the grandest things humans have ever conceived of doing, and burdening it with petty terrestrial agendas.

Here I am at 43 and we are still flying the same Space Shuttle I saw lift off in 1981 at the age of 26 - a spaceship designed in the early 1970's when I was a teenager. And now this relic is going to dock with a new space station we could have built 20 years ago. This is not progress. Nor is this exploration. Rather, this is stagnation - indeed, it is willful abdication from our traditional leadership role in space exploration playing itself out in slow motion.

My generation was robbed of this reality and now another generation may get shortchanged as well.

Problems and Solutions

Unlike many people, I do not see this crux in our exploration of space as being so simplistic as to have been caused by too much government involvement or too little private sector participation. Rather, as mentioned above, I see the dilemma we now face as being due to a lack of a coherent national space policy - one which sets goals, defines priorities, and provides a forward looking plan that people can depend on.

When it is clear who is going to do what, and when, it should be a straightforward process to decide whether public and/or private sector approaches are most appropriate for the tasks ahead. It should be obvious to anyone who watches NASA that it has been burdened with many operational responsibilities which the private sector clearly can perform more economically. On the other hand, there are intrinsic research endeavors that often require decades to demonstrate economic potential. These things tend to be best done by governmental entities.

Right now, we have no clear plan or vision. Absent such a plan, the whole process of exploring and using space has defaulted to competing factions to whom coordination and cooperation provides no obvious benefit. I am often amazed that we get anything done in space!

While I have no idea whether there will be a government agency named "N.A.S.A." in the decades to come, yet I see no reason why there could not be one - or some collection of agencies with similar charters. I could also easily imagine that government's role would be somewhat diminished and that private concerns have expanded out to lead the way much the way it has happened on Earth.
I can predict, however, without a thorough rethinking and overhaul of the way we explore space, we'll see less progress in the future, not more.

As I mentioned earlier, we are all to blame for allowing this dysfunctional method of space exploration to arise and persist. Citing problems is easy. Coming up with solutions is not -- especially if one is required to admit to being part of the problem. What follows are some general problems, paths for solutions, and a look ahead to the importance of solving these problems.

Problem: NASA is not an airline: We're all at fault by transforming NASA from a lean, mission-oriented research, development, and exploration agency into a trucking and communications firm. NASA is at its best when it is pushing technological frontiers and going places no one has ever gone before - propelled by goals and dreams along the way. NASA is at its worst when it tries to perform the routine operations in that frontier more appropriately suited for the private sector.

We should have learned by now to run away screaming from anyone trying to tell you that a government entity can operate any transport service economically. Amtrack anyone? We should, in turn, stop and take heed when government says that some things, such as the search for knowledge, are worth doing regardless of economic payback - and can only be attempted with the resources only a government can amass.

We need a NASA that can say "No" as easily as it says "Yes". We also need a NASA who can tell us "why" something needs to be done in an open and honest fashion.

Solution: The entrepreneurs who run startup firms and begin from zero to grab market share are often ill-equipped to run those very same companies once they have matured and achieved stability in the marketplace. NASA is best at pushing the technological frontiers - not in maintaining them on a daily basis. It is time to get NASA back to what it was chartered to be.

NASA should commercialize routine launch and operations activities as soon as possible. Attempts made thus far are promising but fall far short of what is needed. The FAA pilots do not fly commercial airliners. NASA shouldnÕt put itself in this position either.

NASA should retool itself to focus upon its chartered responsibilities. The Administration and Congress should exercise restraint and not load NASA up with tasks designed to make the constituents happy back home. Rather, they should keep their eye on the prize and let the chips fall where they best suit NASA - not the next election. Although I think that modern telecommunications makes one's physical location increasingly irrelevant, NASA should be allowed to close and open centers as it deems fit in order to accomplish its tasks.

However, when no clear compelling geographical preference exists for a task, NASA should seek out and use excellence regardless of where it is to be found. With people telecommuting everyday from a space station it is going to become increasingly hard for NASA to justify sending people and projects to other centers - unless some unique physical facility warrants such a move.

A Look Ahead: NASA has a lot on its plate: missions are being flung all over the solar system, humans will soon living in space permanently, extensive data sets will come pouring out of skies at ever-increasing rates. It is more than enough to expect NASA to deal with this deluge of scientific data without having to operate all of the logistical systems needed to keep things going. NASA has a chance of thriving so long as it knows when to let go of something and let the private sector make it sink or swim. In he rush to commercialize and privatize routine operations, however, NASA needs to be certain that it does not loose valuable capabilities if there is no portion of the private sector ready and able to accept the task.

Problem: Lack of Presidential Leadership: The President and all of his staff at the White House are at fault for issuing a space policy and then promptly ignoring it. The Administration is also at fault allowing economic and foreign policy interests to manipulate our space program - the most blatant example being the abduction of space station program by the State Department for use as a conduit for foreign aid to Russia.

Moreover, when real problems have arisen with the space station program, ones which require their considered attention, the White House is no where to be found. An Administration whose re-election theme song was "Don't Stop Thinking About Tomorrow" has done just that when it comes to the exploration of space. Such abject avoidance of this problem has been the direct cause of many delays in the space station program since NASA has not been empowered by the White House to use all of the resources at its disposal in order to fix the problems.

Instead of removing Russia from the critical path, the White House has stood by and allowed Russian non-performance to delay progress and drive up costs. Then, to compound this situation, the White House has the gall to tell NASA that these problems all have to be solved within NASA's existing - and shrinking budget. The net result is that all of NASA's other programs are, or will be, delayed, shrunken, or abandoned.

The net result is that the space station program has technical decisions being made for political reasons and political decisions being made for technical reasons. Small wonder this program I such a mess. If this is how America intends to run all future, large space programs, I predict that there will be no such programs.

Solution: It is obvious that the current Administration is unlikely to change its approach to space policy any time soon. Regardless of who is elected in 2000, they would be well advised to reinstate the National Space Council. This Space Council should be populated with individuals who can serve as sources of expertise for the development of a cohesive national space policy and, more importantly, as watchdogs for the implementation of this policy.

A national space policy needs to be completed within the first year of the new Administration. We cannot afford to delay much further. This should include an implementation plan and budget request ready for submittal during the second budgetary submission made by this Administration. Everyone needs to know what America is going to do in space, why we are going to do it, when we are going to do it, and how much it will cost. If we decide not to do something in space, a clear rationale needs to be voiced so as to leave no doubt what is and is not a goal of our nation in space.

Politics will always be a part of the exploration of space. However, when political concerns become programmatic drivers, the Administration needs to address these concerns openly and honestly. If a space project is to have a foreign aid component (as is the case with the International Space Station) then funding for that project needs to draw upon those portions of the budget normally concerned with foreign aid i.e. the State Department. Not to do so is being less than honest with Congress and the public.

A Look Forward: Imagine today's dysfunctional approaches extrapolated onto a program of Martian exploration. If we have so much trouble supporting a multi-national, multi-year multi-billion dollar program in low Earth orbit, imagine what we'd face if we mounted a similarly complicated program to send humans to Mars! What happens if one key member of an international coalition experiences a change in government wherein their contributions shrink or cease? Can the program be designed with enough robustness to proceed? How do we establish mechanisms to manage probes on missions out of this solar system - missions that will require decades of support? We're never going to explore the solar system if we can't take a spin around the local neighborhood.

Problem: NASA's Broken Promises: NASA is at fault for making promises to the White House, Congress, and the taxpayers without first doing its homework. NASA often puts an overly optimistic spin on the true cost of a project and the risks involved for fear of having the project killed as a result of White House and Congressional sticker shock. The International Space Station with its politically motivated Russian participation is the most recent example. Once a project is approved, a creeping cost incrementalism then enters into the equation. The underlying hope at NASA being that somehow money actually needed can be cajoled out of Congress in small chunks so as to fix all of the things that were overlooked during the initial selling of the program. Over the past several decades this approach has been refined into a high art form.

NASA's contracts are too easily awarded, often to a select concentration of contractors among which competition during the proposal process is now more of a hope than an actual fact. Indeed, merger mania has reduced a collection of aerospace contractors down to two monolithic alternatives - both of which also control the Space Shuttle. How anyone can even suggest that competition is enhanced by such an arrangement escapes me.
My only hope is that some of the new methods of getting into and using space may be soon be done by smaller firms who relish their independence and have to compete everyday to keep their share of the market.

When costs start to rise and contractor problems retard progress, NASA comes to Congress to beg for another money fix to get them out of the cost overrun du jour. Requesting transfer authority well into a fiscal year, shifting the phasing of facilities, and moving research grant anniversary dates and other procedural mumbo jumbo are often used to hide the fact that other programs within the agency find their resources being cut to support the Space Station or Space Shuttle. NASA then asks for forgiveness and pledges that it doesn't need any more money only to show up a few weeks later asking for more, new excuse in hand. This process is allowed to go on year after year because the White House and Congress let NASA get away with it.

Solution: NASA itself needs to practice what it preaches in the innumerable project management classes it makes its employees take. NASA needs to spend the required time up front to understand the requirements, technological risks, and funding environment before it proposes a new start - not after it has started the task. Waiving management fads such as TQM and ISO 9000 in front of NASA employees is not going to fix a badly designed program. Instead it just serves to increase their frustration since they know that NASA Headquarters will never adhere to the same principles they hoist upon their subordinates.

When NASA becomes aware of a potential cost increase it should be the one who reaches out to Congress - not the other way around, as is all too often the case. How NASA can expect Congress to help them out after NASA has withheld information until the last moment, again and again, always baffles me. I am even more mystified to hear people at NASA complain that Congress doesn't trust them.

NASA also needs to demand much more responsibility from its contractors. Incentives for contract success should vastly outweigh those for failure. Indeed, right now, the penalty for failure is often rather minimal given the cost plus arrangement favored in most of NASA's large procurements. The worst that can happen when things go awry is that all contract costs are covered and a CEO has to go to the Capitol Hill to endure a verbal lashing, protected by his golden parachute all the while.

A Look ahead: Imagine what resistance a future NASA Administrator will face trying to sell another large program (e.g. mars) to Congress. If NASA's current approach to selling and managing large programs is not fixed rather soon there will be no new large programs in space for decades to come. Indeed, I predict that NASA will not be allowed to mount any new human space programs until such time as space station assembly has been completed, the space station has operated for several years and has begun to deliver some of the numerous benefits NASA claims it will provide.

Problem: Better, Cheaper, Faster works, but not for everyone: NASA touts its "better, cheaper, faster" mantra for the radical change in mission design and operation but clearly does not enforce it in all of its programs. Some sterling examples have emerged from NASA - most notably the Mars Pathfinder mission. This approach also includes tough guidelines wherein programs are cancelled if they stray from pre-set ground rules. The cancellation of the Clark mission being one example of this policy in action.

Given NASA's focus on this approach, you'd think that the preeminent place to practice this new way of doing business would be on the International Space Station program - NASA's largest single activity. You'd think this would be especially true inasmuch as the ISS program was created as a cheaper and better way to build a space station using remnants of the Space Station Freedom Program. What part of the new, improved ISS program conforms to the cheaper-better-faster way of doing business? I certainly cannot find it. It is not "cheaper", it is not "faster", and only time will tell if it is "better".

Solution: NASA needs to be honest up front about the costs and risks associated with any new project. If NASA is going to set standardized criteria for program cancellation it should be ready to apply the same criteria against all of its programs i.e. the International Space Station. The Clark mission was cancelled when its projections showed it running by 20% over its $51 Million expected cost. In February 1998, NASA FY 1999 budget estimates showed the Space Station program exceeding it's advertised developmental cost of $17.4 billion by $4.2 billion for a final expected development cost of $21.66 billion. This represents a 24.5% cost overrun. I am not advocating the cancellation of the Space Station, but I am suggesting that rules that are ignored are often worse than having no rules to begin with.

A look ahead: There is no doubt that NASA's Better, Cheaper, Faster approach has clear, proven merits. But invoking it only when politically desirable is not going to provide all of the benefits a broader policy would provide to NASA. It may also prove to be true that some projects, most notably human space flight, may not be as amenable to this approach as are robotic missions. Only time will tell.

I suspect that as we actually move our robots, and then ourselves, out into the solar system that a more pragmatic, adaptive approach will emerge - one based upon actual experience. For all we know deep space missions may require quadruple dissimilar redundancy as a standard requirement. A look at some of the technological approaches attempted by well-meaning, and well-prepared Antarctic explorers serves to illustrate how wonderful ideas at home may be worthless out on the frontier. Perhaps the phrase "Better, Cheaper, Faster" should include "Smarter".

Problem: Congress as an observer: Congress is at fault when it stands by and allows things to happen. Congress needs to become part of the solution to problems, not just the place where everyone runs with their problems or to complain about others. There also seems to be little in the way of teeth in any measures Congress manages to pass. NASA misbehaves, yet Congress keeps the money flowing.

All too often Congress takes the credit when things go well at NASA, often naming things at NASA (or in their home state) after each other in the process. Yet when things get bad, Congress rushes to point fingers and assign blame - blame which is almost always vectored away from Capitol Hill. Congress also seems ill equipped to listen and respond effectively to bad news when they are not in a mood to hear it.

As an example: the issues facing the ISS program as a result of Russian participation have been known since day one within Congress. Despite a torrent of input from people within NASA about the risks associated with bringing Russia into the ISS program, Congress signed off on the deal.

Now, after half a decade of playing an annual game of brinkmanship driven by a steady series of funding standoffs, Congress, NASA, and the White House are all still blaming each other for the current situation. Each year outraged lawmakers rail against NASA's delays and the White House's indifference, yet each year the margin for votes to keep the space station grows and grows. Where's the incentive upon either NASA or the White House to change their tactics? Congress always gives them what they want in the end - with an ever-growing consensus! From NASA's perspective, they seem to have hit upon a winning formula!

When Congress does deal with these problems, it only does so after they have become larger than they need to be. NASA is at fault as well since it harbors a well-practiced habit of not telling Congress anything they do not feel it needs to know. Instead of giving Congress advance notice of potential problems, everything is allowed to stew and ferment until it bursts out forcing people to take rash actions instead of making well-considered decisions.

I do not want to suggest that Congress has done nothing when problems have arisen at NASA. But it certainly could do more. Congress must accept and exercise the full extent of its oversight responsibility and hold those responsible accountable for not delivering on their promises. If need be, Congress needs to start rolling their sleeves to help when help is what is needed. If this current situation does not change we will never explore the solar system.

A look ahead: Several years ago, as the 1996 presidential election loomed ahead, there was bipartisan support for a "space summit" - an effort wherein America's space priorities would get a re-examination, re-adjustment, and re-validation. Soon after the election was over, interest evaporated. The White House laid an ever increasing series of ground rules upon the summit which would have rendered it a waste of cab fare from Capitol Hill - and supporters found something else to be interested in.

This is a shame inasmuch as the euphoria that had accompanied the discoveries within the ALH84001 Martian meteorite had captivated the public. Why NASA did not urge the White House to use this opportunity (or vice versa!) to substantially bolster NASA's program of planetary exploration still mystifies me. Now we see unnecessary compromises being made to the core science missions in several Mars projects because they have exceeded their cost cap - and there is no money left to spare.

The failure of the space summit was also unfortunate since some of the disconnects between NASA, The White House, and Congress regarding human space flight might have been addressed. Given the stress that ISS delays have had on NASA and its ability to do the other things it is tasked to do, perhaps it is time that this idea be revisited.

Should another attempt be made, however, everything needs to be on the table - budget, priorities, foreign policy, and commercialization policy - the works. And all parties need to abide by a pledge to live with the outcome. Unless some active measures are made to heal these ever widening rifts in regarding the conduct and direction of America's space activities, NASA will become totally obsessed with the space station and generating the cash needed to stall Russia's problems for another week or two. As a result, all of the other amazing things NASA has come to stand for will wither and die.

Problem: NASA lacks an advocate: Instead of being an advocate for his agency's continued health, NASA's Administrator has been a cheerleader for its diminution and has eagerly boasted of budget cuts year after year. If NASA is so hot, then why is its budget being cut when all other science and research activities have enjoyed budget increases? Saying that NASA is better because it "does more with less" only works for a couple of fiscal years before rhetoric begins to conflict with reality. The cuts have gone too far and NASA's fundamental health has been undermined.

NASA's Administrator has the moral and ethical obligation to his employees, his agency, and the taxpayers to make this clear to the White House. I see no evidence that he has done so. Rather, he has dutifully done as the President has told him. Serving at the President's pleasure does not mean that one should not stand on principle when circumstances merit and tell the President the things he needs to hear. If NASA's own Administrator isn't going to stand up for his agency, how can anyone else be expected to?

Solution: NASA's Administrator should be out promoting the merits of his agency, its mission, and its people along with NIH, NSF and other science and technology organizations to the extent allowable under law. These organizations have all witnessed budgetary growth at the same time, and within the same government downsizing environment, that NASA operates within. What are they doing so right that NASA is doing so wrong?

While achieving efficiencies and cutting costs is always a meritorious goal, this should not impede NASA's Administrator from seeking additional work and projects and the funds required to do these projects. Indeed, if NASA is so good at doing things cheaper, better, and faster, there is no reason why it can't tackle larger, more complex missions as a result of having learned these lessons. If NASA's budget is not at least allowed the option of growth at some point, it is simply impossible to even consider any of the grand feats we often envision for ourselves in space - at least not through government funding.

NASA's Administrator should assign himself the action item of totally reconfiguring the way NASA interacts not only with the White House and Congress (the ultimate arbiters of NASA's budget) but also with the general public. Annual Spin Off reports, Public Affairs sanitized press releases, and websites designed to induce sleep are not the way to go.

The exploration of space and the discovery of new and exciting worlds are exciting yet you'd often never know it from NASA. While the visitation numbers have since become almost mythological, the Mars Pathfinder website has probably been visited by everyone on Earth - twice. Has NASA tried to implement lessons learned from that experience?

No. Indeed, the marvelous folks at JPL who jumped for joy on live TV were admonished by NASA Public Affairs for being unprofessional. I'll take jumping geeks in polo shirts over a bunch of well-behaved headquarters nannies in suits any day.

All I can say is the following: when was the last time (other than the 1960's) you saw the press actually cheering for a NASA project ("the little rover that could!")? For a brief moment, that old "can-do" NASA shone forth from Pasadena. Such behavior should be emulated at NASA - not squelched.

A look ahead: It has been difficult enough to get programs in place to perform the initial reconnaissance of our solar system. Clearly if humans are even going to consider moving off of Earth, some rather explicit information is going to need to be available to a large number of people. NASA is going to have to play a pivotal role in public education if we are ever going to see this happen. As such, NASA is going to need to find a way to shed this shyness and tendency to hoard information and get out there and let people know what lies ahead for us.

If NASA itself can't be an overt cheerleader for the exploration of space, then it needs to facilitate the creation of an environment where others can do the task instead. NASA has been given the most amazing charter ever given any government agency. It is shame that NASA cannot utilize every possible avenue to tell everyone what can be done.

Problem: NASA squanders its human resources: NASA's Administrator is at fault for squandering NASA's most valuable asset - its people. NASA has the most inventive, educated, and visionary workforce ever assembled on this planet.

Yet rather than being an advocate for his workforce, Mr. Goldin has gleefully presided over a needlessly harsh and impersonal downsizing process and has avoided even the rudimentary tools successfully used in private industry to treat people in a humane fashion, one which derives from years of dedicated service. Yet once the dust clears from these personnel cuts, Mr. Goldin somehow expects the survivors to feel incentivized to perform better than before. You can't decimate the ranks and then expect to herd the remaining people into productivity with a horsewhip, Mr. Goldin.

Solution: NASA's Administrator needs to abandon his current personnel system and rebuild it from scratch. While the realities of overall government downsizing need to be at the core of the way NASA manages its people, such policies need to go hand in hand with a fundamental commitment to the people themselves. As funds become increasingly scarce, and there are fewer people to do the work, it should be obvious to even the most casual observer that the critical link is the ability of the individual to be supported in all ways as they strive for excellence. As the number of people decreases, the care given to the welfare of each person that remains should increase accordingly.

During my experience running NASA Watch, I have encountered innumerable examples wherein people express their frustration with the way NASA is being run. Yet underneath their current frustration, there is always a resolute commitment to the agency still in place. This commitment to NASA and to the exploration of space may be powerful, but it is not indestructible. As such, my biggest fear for NASA is not budgetary, but rather human.

A look ahead: The time for large personnel cuts at NASA seems to have passed (for now) and NASA's budget seems to have entered into a predictable (although undesirable) state. Now is the time to halt the erosion of human capital at NASA and start rebuilding it. It is also time to start looking ahead to the sorts of people needed to carry out the space exploration of the next century and creating a process whereby those still at NASA can usher in the next generation. To date, NASA's Administrator has paid little more than lip service to this topic. This needs to change immediately otherwise all of the new ways of doing things will never achieve their full potential.

Problem: Irresponsible Contractors: Aerospace contractors are at fault for lining up at the communal trough to gorge themselves with no thought of societal responsibility. A blatant example being on 5 November 1997 when a Vice President of Boeing testified before this committee on cost increases on the Space Station program. Despite the obvious title of the hearing ("International Space Station: Status and Cost Overruns"), he still could not bring himself to say "cost overrun" resorting, instead to the guilt-shifting euphemism of repeatedly describing a "variance" in their " estimate at completion". A $600 Million variance at that. Such performance (or lack thereof) is shameful.

This problem is not confined to human space flight either. One look at how the EOS AM-1 and Landsat programs (both sole source procurements) have been operated and it should be clear that something within NASA's instruction manual for dealing with big contracts is broke. Indeed one of the first pieces of slang I learned when I joined the space station program was "getting well" which as defined to me as what aerospace contractors did to offset the low bids they used to get contracts in the first place.

Solution: Quite simply, aerospace contractors need to wake up to reality and start to show some responsibility and honesty in the way they provide goods and services to the government. If such obscene cost overruns continue to plague NASA the net result may well be that there are fewer programs of such size to bid on. In addition, NASA's Administrator has spoken of adding additional penalties for under performance. The easiest way to avoid making this situation worse is to be open and honest with NASA. It may not be the most savvy business advice, but telling NASA that it is wrong when NASA hasn't done its homework, or that something will cost more than NASA expects it to up front, before the papers are signed, certainly is the ethical thing to do.

A look ahead: Years from now we may well see many aspects of space exploration being done as a private venture with less government participation. This will only happen if the government stops trying to do the operational tasks that private industry is best at doing. However, this activity will likely result only after prudent technological investments by government.

As such, it behooves aerospace contractors to start thinking big picture and long term - and that it may well take a lot more of their own money to "prime the pump" than has been the case in earlier space programs. It may well be beyond their traditional business radar horizon, but any serious expansion outward into the solar system is going to have to rely on more than just 3-5 year business plans.

The aerospace industry has to start to get a little more creative and figure out how an economy spanning multiple worlds is going to operate - and start to factor this into the way they do business - now.

Problem: Reluctant Scientists: Professional scientific organizations are at fault when they allow NASA to use ill-founded hype to sell programs. These organizations are also at fault when detractors of space research release incorrect information and no effort is made to set the record straight.

How many times have you heard members of Congress stand up and talk about the "cures" that NASA promises to develop in space. What "cures" are they talking about? NASA is only doing fundamental research in space. Sure, "cures" could result from this work, but in a decade as a space life scientist, I do not ever recall reading a proposal submitted to NASA wherein the intent was to develop a "cure" for anything.

None the less, do the professional organizations issue press releases to correct these remarks or offer background information on what the facts are? No. Do NASA's life scientists? No. Instead, NASA and the science community sits there quietly and bask in the momentary good news (however incorrect) being thrown out on the airwaves with no thought of the consequences down the road when such predictions don't pan out.

Conversely, when one scientific organization issues a report which attacks the scientific merit of another organization's area of expertise (as the American Society for Cell Biology ASCB did this summer) do other scientific organizations stand up and actively defend themselves, the facts, and NASA's research programs? No, Does NASA take on their critics armed with facts? No. Why? It seems that everyone hopes that people have a short memory and that this will all blow over. The net result: no meaningful dialog about the true potential of space based research ever seems to make its way into a public forum. This is hardly a solid basis to engender support for space based research. All we get is banter composed of sound bites and no follow-through.

The only time you see any overt movement on the part of scientists and engineers is when funding for their particular program or discipline is threatened. Then they howl in protest and trundle a few Nobel laureates up to Congress to set things right. Far too often however, such efforts often include suggestions that someone else's budget get cut in order to preserve another. NASA's science programs will not advance to their ultimate potential if the various factions involved only stand up to support their own disciplines. Support must be forthcoming from all aspects of NASA's science programs in support of all of NASA's science programs.

Solution: Professional organizations whose members are active participants in space based research or who are concerned with quality in government sponsored science need to overhaul their public information efforts. NASA should not be relied upon to be the only path whereby such information makes its way to decision-makers and the public. Members of Congress should be able to avail themselves of honest critiques of NASA's research and the clear benefits that could result from such work.

When organizations or special interest groups attempt to dis-inform lawmakers, the news media, and the public, an effort should be made to take on these activities, and the claims made by detractors point by point. Conversely, when NASA attempts to inflate the scientific value of its programs, professional organizations should be able to rise above petty self-interest and take NASA itself to task. Incorrect information benefits no one.

Moreover, these organizations should expand their outreach beyond the boundaries encompassed by their own members and begin to communication the value of space based research to the general public. NASA needs to enlist the expertise resident within these organizations as it crafts its research and outreach plans.

A look ahead: Picture a website in 2010 dedicated to a space science topic - let's say the search for life on Mars. To me, the ideal website would have a core goal of allowing any visitor to learn as much as they wish to learn about the search for life on Mars - of failures and doubts as well as successes and future plans. It would also let the visitor know how they can participate or support this research.

On that website you should be able to find clear details of all the missions to be performed as evidence of life is sought, when they will be launched, what they are designed to accomplish, what the importance of each mission's task in the context of a larger plan of exploration. On that same website a visitor should be able to locate all of the societies and organizations whose interests overlap with the search for life on Mars. These websites should allow a visitor to tap the expertise of the organization's members as well as learn how to become involved in this activity. This website should also point to the various legislative activities wherein national goals assessed and space research budgets are decided.

Finally, this website should link to sites which take credible issue with the science or the approaches being used so as to allow a visitor to be fully informed to the extent they wish to be informed. They should also openly solicit and post praise and criticism from those who visit and utilize the website. These websites should be developed hand in hand with video, radio, and print materials such that anyone, anywhere can learn whatever they want, whenever they want to. The responsibility for informing all sectors of the public of the value and status of space research is one held jointly by NASA and the scientists and engineers who work on these projects. Neither can do the task alone.

The more informed a citizenry we have, the better decisions all involved will be able to make. The more thorough the decision making process, the more solid the consensus. The more solid the consensus, the more empowered NASA will be.

We are not going to explore the universe with a disinterested and under-informed citizenry. Indeed, I suspect that there is probably some galactic law of evolutionary selection, which does not favor the expansion of an under-informed species out into the cosmos.

Problem: The press and bad news: The press is at fault when it focuses on all the things that go wrong at NASA while spend comparatively little time on what successes are achieved. Most reporters make an honest, consistent attempt to understand the topic they report on. Yet far too often I encounter reporters who are only quasi-informed about NASA and space exploration, often knowing only what is broke and who is to blame, not what works, and who is responsible.

Solution: NASA's somewhat isolationist behavior, high technical content, and peculiar jargon often erect a cultural barrier which reporters are forced to scale - or breach. NASA's Public Affairs Office and the fear of the press they instill in NASA employees does not help the situation either. None the less, this should not prevent a truly good reporter from tying to get the real story even if it takes extra effort. Reporters need to spend more time understanding the NASA peculiarities rather than dismissing them as being impediments. Being a former NASA employee, it is often infuriating to see things thousands of NASA employees do everyday misinterpreted and incorrectly reported by a lazy reporter with a deadline to meet.

With an ever-expanding universe of electronic communications, NASA's ability to hold news in - good or bad - is going to be further eroded. NASA needs to get used to this - soon. Conversely, as evidenced on NASA Watch, the amount of information leaving NASA is going to be in an increasingly raw state begging for interpretation. Owing to the technical nature of what NASA does, one can easily be misled by facts.

In all of my contacts with reporters covering space, I have yet to meet one who, like myself, has reviewed technical and scientific proposals, participated in space station design reviews (regular and Congressionally-mandated), space mission manifest development, annual budget cycles, or payload interface specification development. Yet there is a horde of reporters out there reporting on these topics every week. This scares me. Since so much of what NASA is allowed to do derives from public perceptions, those responsible for shaping those impressions need to know a lot more about that which they profess to understand before they go off and tell people what they think is going on within NASA.

A Look Ahead: Johnson Space Center recently held a workshop for reporters to help them get ready for the International Space Station program. I would hope that this is just the first in a regular series of similar events. NASA should be looking to take this approach for all of its programs. Indeed, NASA ought to consider the journalist's equivalent of Space Camp as a way to enhance the technical background of reporters covering space.

As we move outward into the solar system, the consumption and production of news is certainly going to follow. How is NASA going to handle interviews with astronauts from Mars? How will they read the morning paper? What happens when people on a space station or some planetary base decide that they will handle their own PR?

The expansion of the Internet and the Web across the solar system will be accompanied by the distribution and access anarchy which it seems to elicit from people. I predict that any attempt to regulate the flow of information in either direction will only happen as a result of constraints imposed by bandwidth and the speed of light - not on the basis of content. As such, it is time for NASA and the fourth estate to start figuring out how to make this work to their mutual benefit instead of their mutual distrust.

Problem: Xenophobia at NASA Public Affairs: NASA's Public Affairs Office (PAO) is at fault by virtue of having become the de facto Propaganda Announcement Office with the singular role of preventing the release of damaging information. When bad news does get out, NASA PAO seeks to put the best possible spin on it. For information released voluntarily, NASA is often its worst enemy. I have seen far too many examples of amazing and exciting things NASA does "dumbed-down" for public dissemination. Instead of going out of its way to make the agency open to public scrutiny, NASA PAO seeks to keep the public out.

NASA PAO seems to have a mission focused only on purveying happy, positive thoughts. If you visit their Space Station or Space Shuttle websites, you'll see that they post reader comments. Have you ever seen a comment in anyway critical of NASA? No - nor will you. When NASA put together its 40th Anniversary exhibits of pivotal events in NASA's history, was there any mention of the Apollo 1 or Challenger accidents? No. NASA has become so xenophobic that it is incapable of admitting, much less dealing with any external criticism. Look at the way they craft their congressional testimony and you will get a regular reminder that they just can't admit that they are at fault.

NASA's greatest asset is its employees, civil service and contractor alike. Yet from the way NASA PAO overtly prevents them from acting as ambassadors to the outside world youÕd think they were guilty of some crime. Indeed, recent surveys done by NASA itself show that an overwhelming portion of NASA employees do not feel that they can speak out freely with out fear of retribution.

When NASA contractor employees speak out, the fate is far worse. When Jim Oberg, Ken Hollis, and Tom Hancock (a.k.a. "BitFlip") exercised their constitutional right to free speech, and discussed NASA without PAO permission, they soon found their jobs in jeopardy such that they had to leave their jobs. These individuals spoke of nothing proprietary and often spoke and wrote things that made NASA look good.

Any organization, which is so eager to silence, those who do not agree with official agency dogma is an organization with a serious case of insecurity - one which is not in keeping with the best interests of its employees, its mission, or the taxpayers it is supposed to serve.

Solution: NASA Public Affairs needs an overhaul. This culture of hoarding and manipulating information has to end. Reporters should be viewed not as adversaries but rather, as intelligent individuals who respond to an open honest answer with an open honest story. PAO complains that reporters get their facts wrong yet PAO goes out of its way to shelter them from information as long as they can possibly get away with it. Small wonder that the press corps covering NASA holds the much the same general suspicions about NASA's truthfulness as does Congress.

Given the nature of space missions, NASA is responsible for getting the news from immense distances and bringing it back to Earth. With such a role comes the temptation to control the flow of information. Instead of deciding what NASA wants people to hear, NASA should seek to provide them with what they want to know. When problems loom ahead, NASA should be the one who calls reporters with an alert. All too often, stories, which come out of NASA, get inflated out of proportion for silly reasons.

A Look ahead: The era when newsmakers made news, reporters covered it, and people read what reporters wrote are fading fast. With the advent of various electronic networks, the line between consumers and producers of news is already blurred. Look at what home video recorders have done to local news coverage. Now imagine what the world will be like when all of the low Earth orbit global telecommunications systems become operational. Today you can have a website with global access for $19.95 a month.

If NASA PAO finds NASA Watch a painful annoyance, imagine life in the near future, where a million people around the world decide to sit at their home computers (like I do) and "watch" NASA. If NASA adheres to its current xenophobic mindset and does not adapt to the fact that its information control is going to diminish, NASA will soon find itself increasingly consumed with reacting to events, instead of making them happen.

Problem: NASA employees: Its your agency: Fix it. The employees of NASA are at fault for not demanding better management. Despite the current Administrator's admonition that NASA does not belong to its employees, the last time I checked none of the civil servants or contractors who work for NASA had surrendered any of their rights as taxpaying citizens. This includes their right to communicate their views in public and to their elected representatives. Moreover, as the actual rocket scientists who make the space program work, they should be expected to serve as NASA's best outreach resources.

Alas, it is not so simple to stand up and speak out. Indeed, there is good reason not to speak out at NASA. Recent surveys taken by NASA itself show an overwhelming majority of NASA employees afraid to speak out for fear of recrimination

None the less, the people who make up the NASA family need to get much more active. If the employees of NASA do not care enough to take a risk and be heard, they have no right whatsoever to expect anyone else to do it for them. I am tired of hearing NASA employees complain in private but shy away from even the thought of doing something about their concerns.

A look ahead: Right now it is "career adverse" to speak out at NASA. NASA's Administrator claims to support openness at NASA yet his shoot the messenger policy is still very much in place. Someone else will eventually run NASA. I would hope that this person would seek to overtly reverse this unfortunate policy as soon as they come onboard.

A NASA wherein each and every employee is tasked to be an ambassador not just for NASA, but for the exploration of space, would be a powerful motivating force in our society. It used to be that way back when people didn't know any better than to be openly enthusiastic about working at NASA - and free to say so. I can predict that a NASA wherein this sad state of affairs is not rectified is one doomed to failure.

What vision for America's space program do I see?

As I mentioned before, I am not at all certain what such a vision should be. In setting a vision and goals, care needs to be taken to set them close enough so that people stay focused on the task - and that they exert a personal relevance to all involved. President Bush might as well have picked the year 3000 when he said that American should be on Mars within 50 years. The goals and visions need to have practical as well a scientific relevance. They also need to fill what I call the "yearning component" within all of us. Looking back at Apollo can be illustrative but it can also be deceiving. That was then, this is now.

Personally, I would suggest that America commit itself to finding life elsewhere in the solar system - and then beyond. This will be done at first by robots. Yet I feel that it is inevitable, indeed and desirable that humans show up once the path has been blazed by machines.

Given recent discoveries on Mars, Europa and elsewhere I am all but certain that we'll find it. NASA has recently begun a new research initiative - Astrobiology. In its broadest interpretation, Astrobiology seeks to understand how the universe organizes itself to form life, how life evolves, and how life, in turn, expands outward into the cosmos from its planet of origin. I cannot think of a better organizing theme for NASA as it turns 40. The popular vote is in on this topic. A simple glance at the top box office hits should make that abundantly clear.

NASA at 40 is an agency poised on the cusp of what it was and what it will become. NASA has lead humanity's initial reconnaissance of the solar system and has hurled probes out into interstellar space. We now stand poised ready to take the next great step outward. This time we need to move outward with the full of intent of staying.

While technology does not yet exist to achieve the things we'd like to do, this is not the most difficult hurdle we have to clear. Rather, the hurdle we need to surmount is within ourselves and our ability to rise above the mundane politics of the day and reach for the stars.

Keith Cowing

Keith Cowing is trained as a biologist and has a multidisciplinary background with experience and expertise that ranges from spacecraft payload integration and biomedical peer review to freelance writing and website authoring.

Cowing is editor and webmaster of NASA Watch, an online publication devoted to the free and uncensored exchange of information on space policy and NASA operations. This website is read regularly within NASA, Congress, and the global space community. As the editor of NASA Watch, Cowing has been interviewed by the Newshour with Jim Lehrer (PBS) and the BBC World Service, cited as a news source in Congressional hearings by the chairman of the House Science Committee, and profiled by Government Executive Magazine, Gannett News Service, MSNBC, and Florida Today.

Cowing is also webmaster of The Astrobiology Web, which is designed to be "an online guide to the living universe". The Astrobiology Web has been cited for excellence and utility by a number of broadcast, entertainment, and professional media, including CNN, NPR's Science Friday, OMNI magazine, New Scientist magazine, Scientific American Frontiers, Final Frontier magazine, The National Space Society, The SciFi Channel, and the official website for the recent film, " Contact". In 1998, the Astrobiology Web was formally endorsed by the National Space Society.

Between 1990 and 1993, Cowing was a NASA civil servant and served as Manager of Pressurized Payload Accommodations at the Space Station Freedom Program Office (SSFPO). As part of his prime responsibilities, Cowing served as the Payload Accommodations Manager for the 2.5 Meter Centrifuge Facility, the Gas-Grain Simulation Facility, the Gravitational Biology Facility, and the CELSS Test Facility. Cowing also participated in all SSF design reviews held during 1990-1993, representing the interests of the scientific payload community. Cowing was also the SSFPO lead on biospecimen containment and was a NASA representative to the NRC Committee on Toxicology's subcommittee on Spacecraft Maximum Allowable Concentrations.

Prior to his work at NASA, Cowing worked for the American Institute of Biological Sciences (AIBS) on NASA Life Sciences Division Peer review. Cowing also worked for the Universities Space Research Association (USRA) where he managed a subcontract in support of the NASA Life Sciences Strategic Planning Study Committee (a.k.a. the Robbins Committee). Cowing is also a former member of the governing board of the American Society for Gravitational and Space Biology (ASGSB).

After leaving NASA and returning to AIBS, Cowing managed various peer review contracts, including the FY 94 and FY 95 peer reviews for NASA's Office of Life and Microgravity Science and Applications. Keith also had a senior management role in the AIBS review of 2,668 proposals for the Army's 1994 Breast Cancer Research Project, the largest single peer review project ever performed by or for the Federal government.

Cowing has had several other "careers". He has worked as a certified sign language interpreter for the hearing impaired at several universities and signed various speeches from the podium of the 1980 Democratic National Convention; was an advance man, local organizer, and finance officer for Jerry Brown's 1980 Presidential campaign and the 1982 Brown for U.S. Senate campaign; taught biology and sign language on several college faculties; and worked on the Space Shuttle Program at Rockwell International's Downey.

SPACE FOR AMERICA

The following professional organizations and institutions have endorsed this statement:

Aerospace Industries Association
Aerospace Medical Association
Aerospace States Association
American Astronautical Society
American Institute of Aeronautics and Astronautics
American Society of 'civil Engineers
Center for Engineering Infrastructure and Sciences in Space, Colorado State Univ.
Challenger Center for Space Science Education
East-West Space Science Center, University of Maryland
Environmental Research Institute of Michigan
IEEE - Aerospace and Electronics Systems Society
Institute for the Social Science Studies of Space
NASA Alumni League
National Space Society
Space Transportation Association
The U.S. Committee for the International Space University
United Societies in Space
United States Space Foundation
Women in Aerospace

Summary

This statement, endorsed by the space related professional organizations and institutions listed above, was prepared by the National Committee on Space. The Committee, established by the American Astronautical Society in February 1995, is composed of representatives from the various professional organizations. What follows stems from concerns the organizations share about the future of one of our nation's premier activities: the exploration and use of space.

Thirty eight years ago, humans placed an object in space, heralding the Space Age and opening a new frontier. This single epochal event ignited the imagination of people throughout the world and led to a cornucopia of great discoveries and achievements in the decades that followed. Humans soared into space, explored the Moon, probed the depths of the universe, and employed technologies developed for space achievements to provide greater security, better health, higher education, and higher quality of life on Earth. Recognizing the vast potential of this unique treasure, the American people strongly supported the exploration, development, and use of space, while enjoying its fruits.

Now, However, the world has entered an era that has necessitated a sea change in the nation's priorities for dealing with new realities. Most federal budgets, including for civil space, are being cut to reduce the federal budget deficit. The implications of a steadily declining public investment in space over the years could be profound and far reaching. Overall space activity will slow down, the space industrial base will shrink, and the momentum of the progress being made will be lost. The nation's ability to meet its future needs could be seriously weakened.

Space is about the future of our nation and humankind. It is a frontier with unique attributes and great promise. It provides a multiplicity of benefits in many critical areas. Publicly and privately supported, space activities help us enhance our nation's intellectual and economic leadership position in world affairs. Space exploration and science help us understand the universe, the solar system, and the planet on which we live. Observations from space help us uniquely in managing our planet's limited resources and fragile environment, as the world's population and needs grow. Space activities help us enhance national and international security. And space activities in general help us improve our quality of life and enrich the scientific and technical education of our youth. Humans have expanded their horizons continually throughout history; we believe they will expand into the frontier of space and over time establish settlements in it and use its resources for their well-being on Earth and in space.

We believe that our nation should pursue the following vision over the next century:

> **A strong America lending the space efforts for humans to live, work, and enjoy a secure, high quality of life on Earth and in space.**

To realize this vision, we recommend that our nation work towards the following near- and far-term goals:

- Affordable space transportation and operations
- Vigorous economic activity in low-Earth orbit and beyond
- Broader knowledge of space and the universe
- Deeper understanding of the Earth and its environment
- Increased understanding of the human as an inhabitant in space
- Expanded education related to space
- Enhanced security and protection
- Human settlements in space
- Broad international cooperation

Affordable space transportation is fundamental to all space activities. It is the most urgent of all goals and must be reached expeditiously. The cost of space transportation and operations must be reduced by at least a factor of ten. To develop new technologies and encourage growth in the transportation industry, the federal government should continue joint efforts with industry, purchase transportation services from industry, and transfer routine operations to the private sector.

To encourage vigorous economic activity in space and involving expanded practical application of space technology, the federal government and state governments (where appropriate) should: collaborate with the private sector in developing generic space technologies; expand efforts to make space technology readily available to private sector for the development of economically promising applications; minimize policy, regulatory, and administrative constraints, restrictions, and burdens on the private sector; and provide financial incentives. Industry should rely less on public funding and more on private investments to develop space for economic activities.

To advance our knowledge of space and the universe, the nation should continue to conduct vigorous space exploration and science activities. The nation should deploy advanced sensors in space to study the universe; send precursor, robotic missions to the moon and Mars, followed by international human missions to Mars in the second decade of the next century; establish a permanent international lunar base; and search for other planetary systems and life beyond the Earth.

To increase our understanding of the Earth and its environment and to encourage growth in remote sensing industry, the federal government should continue to support studies of Earth's environment and climate; develop environment and resource monitoring technologies and systems; lower regulatory barriers to expedite commercialization of technology; become a buyer and not a supplier or seller of remote sensing information; and allow the use of information derived from the national security space systems to monitor the Earth's resources and environment.

To increase our understanding of the human as an inhabitant in space, the federal government should conduct a vigorous life science program with emphasis on medical, psychological, and social factors affecting human performance in space.

To enhance education, both the government and industry should boost space related educational activities. Concurrently, educational institutions and professional societies should conduct comprehensive, space-related education programs at various academic levels; promote advancement of space science and technology by holding professional symposia and giving awards and scholarships; and conduct space education outreach programs for both students and the public. The media should be encouraged to run more space-related news and educational programs in their publications and broadcasts, focusing on public interest and value of space.

To enhance security and protection, the federal government should continue its defense-related activities at the required pace, seeking international participation for international security. Also, the government should explore ways to protect the Earth from possible collisions with large asteroids.

To aid the evolutionary process of expansion into and eventual settlement of space, the nations should continue to develop the required knowledge base and key technologies, initiate international efforts where necessary, and seed international cooperation on establishing a legal code on property rights in space.

Our nation should seek broad international participation in as many areas of space activities as possible. International cooperation is essential to achieving the vision we see for America.

Introduction

This statement was prepared by the National Committee on Space and endorsed by the various professional organizations and institutions listed in the preface. The statement is intended to assist national leaders in setting a strategic course for the nation's space activities over the next century and to show why the nation should continue to conduct a growing space activity and what the nation should do to help meet its future needs. The audience for the statement encompasses federal and state governments, industry, academia, media, and the general public.

Serious concerns led to the development of the statement at this juncture. Civil space activity in our country is currently being funded predominantly by the public sector and to a modest extent by the private sector. While private investment in space activities has been growing, public investment in space is slated to decline steadily to help reduce the federal budget deficit over the next few years. Projections show that by the year 2000, the funding would be reduced by approximately forty percent.

The implications of the projected decline in public investment in space could be far reaching. Over time, the resulting contraction in space activity could seriously affect the future of our nation and its role in the new world order. It could erode the nation's ability to meet its future needs, to compete with other nations in the global market, and to sustain its intellectual and economic leadership in world affairs. Space activity is of tremendous strategic importance to our nation; its contributions are multi-faceted and unique; its highly visible achievements symbolize the strength of our nation. The activity, however, must have stable support if it is to continue to make such contributions and thereby sustain the strength and well-being of our nation.

This statement contains: (a) our **observations** on the contributions of space, on the changes that have occurred in the global environment and in national priorities, and on what is required in space activities to help meet the nation's needs in the next century; (b) our underlying **beliefs** pertaining to space and the future of humanity, and (c) our **recommendations** on future directions for the nation's space activities.

Observations

There is little doubt that over the relatively short period of 38 years, space activities throughout the world have contributed significantly to the advancement of humanity. The activities helped us understand our universe and meet our basic needs for security, economic growth, health, and quality of life. For the first time in human history, we have developed the means to free ourselves from the bondage of gravity, explore the universe, and expand into it. We have launched thousands of satellites into space, sent humans to the moon and probes to the planets, deployed a laboratory and a station in space, studied the effects of space flight on human health, transported humans and machines routinely into space, and sent powerful instruments to probe and understand the universe. These scientific and technological achievements have captured the imagination of the world and ignited the interest of our youth in space, engineering, mathematics, and science. Space has also helped us to meet foreign policy objectives and to maintain peace and security in an increasingly tense world.

While continuing to satisfy our quest for exploration and understanding of space, we have used space technology to meet a variety of our needs on Earth. Today, we are using space systems to provide security, to communicate and navigate around the globe, to advance medical technology, and to manage our planet's natural resources and environment. Investments in space have led to new industries -- in satellite communications, satellite launch services, satellite navigation, medical instrumentation, commercial robotics, and remote sensing. We are now building an international space station, in partnership with other nations, to learn how to live and work in space and how to use it to meet our needs on Earth. We are paving the way to travel and settle in space and to extract further social and economic benefits from it. However, we have not yet tapped the full potential of space to satisfy our growing needs.

In the United States, space activities helped the country significantly in maintaining its leadership position among nations. These activities, however, were driven largely by security concerns and the technological and scientific competition between the two superpowers. The federal space budget rose and fell accordingly over the years in response to competition and the threat. The civil space budget rose sharply in the 1960's primarily to win the race to the moon; the defense space budget peaked in 1988 along with the overall defense build-up of the 1980's.

We have now entered a new era. The Cold War that was driving the space activity is over. Our erstwhile rival, the Soviet Union, is survived by several new republics. Cold War competition has been replaced by global economic competition. While mindful of national security, our nation's attention is now directed inward, focusing on concerns for the economy, the national debt, health, social issues, and the environment.

These new concerns have led to new priorities. Preservation of the nation's economic competitiveness has become a high priority. The need to balance the budget and reduce the deficit has become urgent. Cuts in federal spending have become inevitable. The space budgets -- particularly the civil portion -- are being cut sharply, putting pressure on NASA to shrink its infrastructure and reduce or terminate its programs. Declining space budgets could tear the very fabric of the nation's civil space activity, set back progress being made in exploration and use of space, halt the flow of space technology to terrestrial applications, shrink the space industrial base, and threaten the nation's near- and long-term intellectual and economic leadership in world affairs.

If the nation is to meet new challenges, preserve its hard-earned position in the world order, and continue enjoying the growing benefits of space technology, space activities must be conducted vigorously with stable public funding and growing private investments. A clear national strategic vision must be declared to guide the activities and sustain public support for them. A series of goals must be established to develop programs and measure their progress. All activities must be in support of the declared vision and the established goals.

To strengthen the national space activity, it is time to rely less on funding from government and more on private investment. It is time to form innovative partnerships among the federal government, local and state governments, industry, and academia. It is time to institute supportive laws, policies, and international treaties to encourage private investment. It is time to adopt new organizational and operational approaches that will improve efficiencies, reduce costs, and maximize return on investments. No private investments will come forth unless opportunities exist to profit from space activities.

Fuller development and use of space for intellectual, economic, or other purposes will be difficult without the pursuit of space exploration and science. History has shown that nations that cease to explore or dare not take risks eventually fall behind. Exploration and scientific research generate knowledge that is essential for the development and use of space. Without exploration, we will not, for example, understand the nature of space and the universe, what resources exist in space, how to use those resources, or where and how to establish human settlements in space. Without science, it will be harder to develop and use space for further benefit of humankind. Space exploration and science also help us maintain scientific and technological leadership of our nation, capture the imagination of the public, and inspire generations to pursue education and research in math, science, and engineering.

Notwithstanding the end of the Cold War, national defense, international security, and long-term well-being of humanity will continue to be of deep concern to all nations on Earth. Space, with its unique attributes, can continue to contribute greatly to the maintenance of peace and security on Earth.

Throughout the history of human evolution, humans have expanded their horizons on Earth. Such evolutionary human expansion can be expected to continue into the frontier of space.

Some space activities will exact a price that no one nation could afford. In such activities, international partnerships must be formed to share costs, risks, and benefits. Partnerships will also help build the climate for international friendship, peace, and prosperity.

The public is the ultimate beneficiary and customer of all space activities. The public is also the principal investor and supporter of these activities. It is therefore vital that space activities address the public's concerns, needs, and desires. Surveys have shown that the public has an innate interest in the adventure and benefits of space; many people have expressed a desire to travel in space as tourists.

While ours is a period filled with challenges, it also presents an historic opportunity for America to solidify its position as a leader and trusted partner among nations, guiding the future of humanity and our home planet, aiding all in living, working, and enjoying a high quality of life securely and peacefully on Earth and in space.

Beliefs

We believe that

- Space is about the present as well as the future of the human species. It is about security, prosperity, knowledge, health, and the quality of life. It is about the well-being of the present generation, as well as opportunities and hopes for future generations. It holds vast potential to help humanity meet its growing needs.
- Space activity is of great strategic importance to America and the world. It is essential for national security and economic growth, for managing our planet's resources and environment, and for understanding the universe and humanity's place within it.
- Space exploration and science are fundamental to the understanding and use of space for the benefit of humanity.
- People are destined to travel into space, engage in increased economic activities in space, and some day establish settlements in space.
- Medical activities in space contribute uniquely to improved diagnostic, therapeutic, and preventive advances for the health of humankind.
- The world will continue to benefit greatly from practical, economically rewarding applications of technologies developed for space activities.
- Wherever necessary, parties involved in space activities -- governments, industry, academia, and private investors -- must join hands in forming national and international partnerships to tap the full potential of space.
- A strongly supportive policy and regulatory environment must exist to encourage the private sector to invest in space activities and apply technologies for economic and other human benefits.
- The public is the ultimate beneficiary, the customer, and the stakeholder in space activities. Its support must be built and sustained by strong leadership, by doing what is in the public interest, by producing concrete results and positive rates of return on investments, and by ensuring that the public is fully informed about the benefits that accrue from space activities.
- Education is fundamental to the achievement of any vision in space and, conversely, space activity stimulates science and engineering education.

Recommendations

Vision

Based on our observations and beliefs, we see the following strategic vision for our nation's space activity over the next century:

A strong America leading the space efforts for humans to live, work, and enjoy a secure, high quality of life on Earth and in space.

Goals

To realize the above vision, we recommend that the nation pursue the following near- and far-term goals:

- Affordable space transportation and operations
- Vigorous economic activity in low-Earth orbit and beyond
- Broader knowledge of space and the universe
- Deeper understanding of the Earth and its environment
- Increased understanding of the human as an inhabitant in space
- Expanded education related to space
- Enhanced security and protection
- Human settlements in space
- Broad international cooperation

These goals span the next century; some of them are short-term goals, others are continuing or long-term goals. Affordable space transportation and operations is the most urgent goal; it must be achieved as soon as possible; without it, it will be difficult to achieve the other goals and realize the vision. Establishing human settlements in space, for now, would be a long-term goal. The rest would be continuing goals with increasing levels of achievement and varying time frames and priorities depending upon changes in global circumstances and national priorities.

Affordable Space Transportation and Operations

Space transportation is critical to all space activities; it must also be affordable to both public and private sectors. It would be impractical to go very far in exploiting space without affordable transportation. Today, the cost of space transportation is too high; it must be reduced by a factor of ten or more. If not, investment costs in most cases will heavily outweigh benefits and only the most compelling space activities will be financially viable. The federal government must commit itself to funding and supporting the research and technology development necessary to achieve this goal.

Although the problem of the cost of space transportation has long been recognized, solutions have been long in coming and difficult to find. Industry and the federal government recently began joint work on three broad programs to develop technologies for a new generation of reliable, low-cost space transportation systems. Such efforts should be continued until solutions are found to this critical problem . The cost of space operations also continues to be high and must be reduced through streamlined operations and improved efficiencies. It could also come down with the advent of lower-cost space transportation and increased space activities.

In order to vitalize and encourage growth in the space launch industry, the federal and state governments (where appropriate) should:

- Continue to develop space transportation technology with industry.
- Purchase space transportation services from industry.
- Complete the transfer of routine launch operations to the private sector.
- Provide financial incentives for the development of private space transportation infrastructure.

Vigorous Economic Activity in Low-Earth Orbit and Beyond

In the new century, vigorous economic activity in space could contribute significantly to the nation's economy and competitive strength. A high level of economic activity could also help generate and sustain support by the public, government and business leaders, and private investors.

The beginning of public investment in the use of space for economic and other purposes can be traced to the "Special Message to the Congress on Urgent National Needs" President Kennedy sent on May 25, 1961, in which he asked Congress to provide $50 million to accelerate "the use of space satellites for world-wide communications" and $75 million to "give us at the earliest possible time a satellite system for world-wide weather observation."

Today, global communications and Earth observation have become major private sector activities contributing to the nation's economy and quality of life. Satellite communications now is a multi-billion dollar industry, and remote sensing is developing rapidly. Both of these activities benefit a long and growing list of industries and human endeavors, including education. All over the world the private sector is making large investments in both industries, with America leading the efforts in most cases.

In the last two decades our military deployed the Global Positioning System to provide navigation support to the fighting forces. Today, that system is being used widely by the private sector throughout the globe for navigational and positioning purposes. A new sub-industry was born and jobs are being created, with America once again leading.

These examples show that public investments in space have led to growth in the nation's economy and leadership strength. In addition, numerous technologies developed for space activities in the public sector have been successfully transfered to the private sector for commercial applications. There are other areas in which economic activity could expand. They include use of the international space station, space travel, and energy from space. The space station could be turned into a "Business Park," offering financial and legal incentives for the development and manufacturing of products in space. Research in space will help us understand the effects of gravity on biological, physical, and chemical processes and develop materials that could not be developed on Earth's surface. Space travel could be developed for the general public, leading to a space transport industry. Solar energy could be transmitted from space to Earth, leading to the creation of new power utilities and less reliance on fossil fuels.

To encourage growth in economic activities, the federal and state governments (where appropriate) should:

- Join the private sector in financial partnerships in developing high risk, high pay-off generic space technologies for economic activities.
- Transfer routine government space operations to the private sector to the greatest degree practical.
- Provide industry and state governments with broader access to space infrastructure.
- Minimize policy, regulatory, and administrative barriers and burdens on the private sector, while recognizing the need to protect legitimate public interests.
- Provide incentives such as tax credits, tax exemptions, and vouchers for space-based economic development activities.
- Lower security barriers to facilitate commercial use of technologies developed for the military, while recognizing the need to protect legitimate national security interests.
- Expand efforts to make space technology readily available to private enterprise for the development of economically promising applications.
- Revisit development of power from space as technologies advance.

Broader Knowledge of Space and the Universe

Our national investment in space exploration over the years has paid off handsomely. We gained more knowledge about our solar system and the universe in the last three decades than in all other ages of human history combined. The knowledge has humbled, awed, and inspired us. The achievements have captured the imagination of our youth and the general public.

The nation should remain committed to space exploration, pursuing programs such as the Discovery program by sending smaller, more frequent missions to the planets, their moons, and the asteroids in our solar system and beyond. Knowledge gained from these missions will allow us to further understand the universe, the origin and fate of our solar system, the resources that lie in it, how we might use those resources, and how we might expand into space.

To enhance our knowledge of space and the universe, the nation should:

- Continue to deploy and maintain advanced sensors and telescopes in space to study the universe and better prepare for scientific opportunities on bodies in space.

- Send precursor robotic missions to study the moon and Mars (including sample returns), followed by international human missions to Mars beginning in the second decade of the next century.
- Establish a permanent lunar base in concert with other space-faring nations.
- Understand the role of fractional gravity in biological, physical, and chemical processes.
- Search for other planetary systems and life beyond Earth.
- Lead in forming international partnerships wherever necessary to achieve these goals.

Deeper Understanding of the Earth and its Environment

The future of the human species and its quality of life is critically dependent upon the health and wealth of the planet Earth. Growing population and industrialization are putting increasing pressure on Earth's environment and resources. Our species will perish if we exhaust our planet's resources, poison its environment, and find no place else to go. Space, with its unique attributes, could help us avoid such a calamitous end. It can help us manage our resources and protect our environment. It can help us inventory our resources and plan for their optimum use and conservation. It can help us monitor the effects of growing industrialization on the environment and the climate (including global warming and ozone depletion) and suggest actions to minimize any damage.

As it did in the areas of communications and navigation, our nation has developed impressive technologies to observe the surface of the Earth and its environment. The nation should remain committed to further development and use of these technologies. The private sector has begun to commercialize land sensing technologies, as the federal government develops technologies for environment monitoring.

To encourage increased use of space to enhance our knowledge of the Earth's resources, environment, and climate, the federal government should:

- Continue to develop environment monitoring technologies and systems.
- Seek international and commercial participation to share development costs and benefits of the environment monitoring program.
- Lower regulatory barriers and offer incentives to encourage the private sector to commercialize both the Earth and environment monitoring technologies.
- Become a buyer and not a supplier or seller of remote sensing information where that information can be generated commercially.
- Allow use of information derived from intelligence satellites to understand the Earth and its environment.

Increased Understanding of the Human as an Inhabitant in Space

Tapping the full potential of space will require increased understanding of the human as an inhabitant in space. This understanding will help us plan and conduct space travel and long-duration space activities with confidence.

53-067 99 - 5

Much remains to be learned about the ability of humans to adapt to long-term residence in space. We do not yet have a full understanding of the biological and physiological adaptation of the human organism to the space environment, nor do we fully understand the behavioral and psychological effects on multi-cultural crews working in space.

Research conducted in space will expand our understanding of the medical and social issues, including the manner in which humans live and work together most productively in space. The insights gained will also contribute to general human health.

To ensure that space travel and long-duration human space missions are planned and conducted with confidence, the federal government should:

- Maintain and expand a vigorous program in basic life sciences with emphasis on environmental factors in space travel.
- Conduct research on social and psychological factors affecting human performance in highly structured work groups.
- Develop cost-effective and regenerative life support systems.
- Ensure efficient transfer of technology and research findings to address medical and societal problems on Earth.

Expanded Education Related to Space

Education of the youth of America and the education of the general public are both fundamental to achieving the vision and goals in space. Education -- particularly scientific and technical education -- of the youth is crucial to the future of our society. Space-related education plays a key role in inspiring, motivating, and attracting young people to pursue careers in science and engineering and in providing the skilled labor force for space activities. Education of the general public is essential to enhancing its appreciation of space achievements and benefits and to sustaining its support for space activities.

National and state governments, industry, educational institutions, professional societies, and the media should all give high priority to space related education and increase their support to those activities. Educational institutions and professional societies should develop and conduct comprehensive, space-related education programs at various academic and professional levels. The professional societies should encourage space related education and research by holding professional symposia and by recognizing professional achievements through awards and scholarships. The societies should develop space related educational programs to attract increased funding from governments and industry. The societies should conduct broad space education outreach programs and encourage the media to run more space-related news and educational programs in their publications and broadcasts aimed at youth and the general public.

Enhanced Security and Protection

In addition to economic security, our capabilities in space contribute uniquely to national defense and international security. The military uses a broad array of space systems including surveillance, navigation, weather, and communications systems in its operations. It also funds research and development of advanced space technologies. As security will continue to be a high national priority -- notwithstanding the end of the Cold War -- these activities should be continued at the required pace. Many of the systems and technologies developed for these activities should be used to derive further scientific and economic benefits for our country. Wherever possible, security barriers between military and civil space activities should be lowered to facilitate the transfer of technology. Broad international participation should be sought to maintain international security through use of space systems.

Space technology may also, some day, help us protect our planet from potential cataclysms caused by extraterrestrial bodies. Millions of asteroids ranging in size from specks of dust to several hundred miles in diameter orbit the Sun in a region between the orbits of Mars and Jupiter. A stream of small asteroids also continuously bombards the Earth, but few ever reach the Earth's surface. Several large asteroids, however, have survived their fiery entries through the atmosphere and struck the surface, leaving huge craters and destroying life. One such asteroid may have been responsible for the extinction of the dinosaurs. The federal government should undertake a project to study the asteroids and identify possible means to protect humans from any threat from them.

Human Settlements in Space

Throughout the history of human evolution, humans have continuously expanded their domain on Earth, discovering new lands and establishing new settlements. We have now discovered the frontier of space beyond the confines of the Earth and are drawn to expand into it. Virtually boundless, the new frontier offers great opportunities for us to expand into space and use its resources to meet our future needs.

Expansion and settlement in space will require significant advances in many technical areas: understanding of the human as a traveler and inhabitant in space, in-depth knowledge of places to settle, efficient and affordable space transportation, long-duration life sustaining habitats, adequate food and power supplies, and the ability to use in situ resources and produce required goods and services.

To aid the evolutionary process of human expansion and settlement in space, the nation should:

- Set human settlements in space as one of its strategic goals and methodically coordinate its space activities towards that goal.
- Lead or participate in international efforts to develop the required knowledge and tools for expansion and settlement.
- Seek international cooperation on establishing a legal code on property rights in space to encourage the pioneering spirit.

Broad International Cooperation

International cooperation in space is critical to the realization of the stated vision and goals. It provides numerous important tangible and intangible benefits. It helps share the costs, risks, and benefits of space activities; it helps bind humanity into a world community; and it helps enhance world peace and prosperity. The United States, therefore, should continue to actively seek international cooperation in as many space activities as possible.

Emphasis should be placed on activities that are beyond the means of any one nation, such as human space exploration and science. One of the first steps toward this goal was taken over 20 years ago, with the Apollo-Soyuz mission. In a unique way, the Apollo-Soyuz initiative helped thaw the Cold War. Recently, we successfully completed the Shuttle-Mir missions. Today, we are building an international space station. Soon, we should move toward the next goals. We should establish a permanent international base on the moon, where scientists from all nations could conduct research and find ways to use lunar and other extraterrestrial resources for use on Earth. The next step should be an international human mission to Mars.

Finally, the U. S. government should take the initiative in developing international rules and space law that will encourage economic development of space for the benefit of humankind.

Testimony to the Written Record

from

Ms. Pat Dasch

Executive Director

The National Space Society

"NASA at 40: What kind of space agency does America need for the 21st Century?"

October 1, 1998

Chairman Rohrabacher, Ranking Member Bart Gorton, and other members of the Subcommittee on Space and Aeronautics, thank you for the opportunity to present testimony on behalf of the National Space Society. On NASA's 40th anniversary it is appropriate to reassess the agency's goals and structure, and to look ahead to ensure it is best poised to meet the needs of the United States as we move into the next millennium.

The world has changed dramatically since NASA was created in July of 1958. The Cold War has ended. Space exploration no longer is the sole domain of the United States and Soviet Union, but a venture shared by many nations. Commercial space expenditures greatly exceed government investments. And the world is rapidly moving toward a global marketplace in which the manufacturing of high technology goods is becoming increasingly competitive and critical to the economic health of our nation.

Sputnik and the New Space Age

NASA is a product of both the Cold War and the advent of the Space Age. The United States established the agency to respond to the Soviet Union's burgeoning space program. But NASA also was created to open the space frontier. Officials realized the ability to launch spacecraft beyond Earth's atmosphere portended a new era of exploration and discovery, and they believed a civilian government agency was necessary to attract the support and coordinate the activities of the science community, which would lead the way.

On October 4, 1957, Russia placed into orbit a 183-pound satellite named Sputnik I. Less than a month later, it lofted a 1,120-pound payload to space, Sputnik II, carrying the dog, Laika. The United States reacted with alarm. According to Dr. Vannevar Bush, chairman of MIT, "it was a rude awakening." It generated "a great shock to the country to learn that in a field where we thought we were doing well, we have been exceeded by [Russia's] performance," he explained in testimony before Congress. U.S. Air Force Lt. General D. L. Putt bluntly declared, "If liberty, freedom, and justice are to prevail for all people, we cannot permit the dominance of space by those who say they will bury the United States." Lt. General James H. Doolittle warned "the United States cannot let space go by default to the Soviet Union."

Problems worsened on December 6, when a U.S. Vanguard rocket exploded seconds after launch from Cape Canaveral. On the world stage, the United States suffered an embarrassing debacle. Congress and the Eisenhower Administration scrambled to find out what had gone wrong with the nation's space program and to implement corrective actions.

"The immediate problem is obvious," explained Rep. McCormack, chairman of the House Select Committee on Astronautics and Space Exploration. "We see another nation of great potentiality, militant and competitive, which has already made the first advances in the mastery of outer space. We cannot stand by and watch this nation make that mastery complete."

"With the launching of Sputniks I and II, and with the information at hand of Russia's strength, our supremacy and even our equality has been challenged," stated then Senator Lyndon Johnson, chairman of the Armed Services subcommittee inquiring into satellite and missile programs. "We must meet this challenge quickly and effectively in all its aspects."

While the Soviet Union's space program posed a daunting challenge to America's national security interests, considered of greater significance was the nation's transition to the Space Age. "There is something much bigger and more important in front of us than a few pieces of military hardware," Lyndon Johnson declared. Ballistic missiles, he argued, "do not hold the key of our future." Instead, he said it " is a new frontier – a new age – that is exciting and challenging." McCormack echoed this viewpoint, stating: "[W]e must enact legislation for the immediate future and must also lay the groundwork for a long-term effort of exploration and scientific development." Looking beyond the horizon, McCormack said, "we are beginning an era of discovery literally as far-reaching as the discovery of our own continent."

There was broad agreement among Washington officials that a new civilian agency should lead America's space program, and not the Department of Defense, which had let the Soviet Union eclipse our nation's space effort. The United States could have launched a satellite to space as much as a year earlier than Russia, but internal rivalry in the Pentagon scuttled the opportunity. In September of 1956, the Defense Department ordered a halt to the work on the Jupiter-C rocket. "After we had fired the first Jupiter-C over a range of 3,300 miles," Wernher von Braun testified before Congress, "we had hoped that this would persuade the Pentagon to give us permission to try a satellite launching, but this hope never materialized." Braun added that "specifically we were not permitted to fire into orbit."

Frederick L. Hovde, Chairman of the Army Scientific Advisory Panel, accused the Defense Department of mismanaging the space program. "The problem before the nation is not simply one of money, facilities, nor even men – these we have in substantial amounts and that which we don't have can be provided," he said. "Whatever the failures may be, they are primarily those of management... which result in delays in decision making and confusion in the direction of our technological forces." The Vanguard program had fallen short according to John P. Hagen, director of the Vanguard Project at the Naval Research Laboratory, because of "too many administrative channels and levels to go through in order to get policy decisions and funds in the time limits with which we were dealing."

NASA's Birth

The main question before Congress and the Eisenhower Administration, as expressed by then Vice President Richard, was: "What type of government agency should have the primary responsibility in the development of our outer space program?" President Eisenhower strongly advocated the establishment of a civilian agency. "I have reached this conclusion," he explained, "because space exploration holds the promise of adding importantly to our knowledge of the Earth, the solar system, and the universe, and because it is of great importance to have the fullest cooperation of the scientific community at home and abroad in moving forward in the fields of space science and technology."

The House and Senate concurred and drafted bills to create the National Space and Aeronautics Administration. "The decision to enter into the Space Age is not one the United States can ignore or defer," a House report declared. "Our national survival requires it." Senator Clark called the creation of NASA "one of the most important activities which has confronted the government of the United States in the history of the republic."

The real purpose of the new agency, observed Senator Carroll, was "to draw into play or operation the scientific talent of the nation...." Maintaining the space program in the hands of the military would have limited the participation of civilian scientists. Lyndon Johnson agreed, adding, "we must draw upon every available source and use it to the fullest extent in order to make the great effort which needs to be made, and there must be coordination of the effort."

The mission of the new space agency was largely defined by the Eisenhower Administration. Its overarching goal is "the expansion of human knowledge of outer space and the use of space technology for scientific inquiry." Reflecting concern about the Soviet space program, NASA was also given the responsibility of preserving "the role of the United States as a leader in aeronautical and space science and technology." A road map with specific projects was left undefined, in large part because officials were uncertain what the future might bring. "The sum of our understanding is not yet sufficient for us to comprehend how vast are the dimensions of our ignorance," a Senate report declared. "We have no frame of reference by which to visualize the design of tomorrow." Senator Saltonstall repeated this view during a speech on the Senate floor. "[I]t seems to me that there are no experts in the subject with which we are now concerned.... All that anyone can do is to speculate about space and its implications for man."

Congressman McCormack mused, "What we will learn from the Moon, and probably the outer planets, no man can rightly say. On the basis of what we already know, we can predict that the advances will be literally beyond our present understanding."

Getting Started

Military space science programs were placed under the management of the new space agency. It also absorbed a group of research institutes. Over the next three years, NASA launched nearly a dozen satellites as part of the Explorer program, and used Vanguard rockets to send science payloads beyond Earth orbit to measure solar radiation, the Earth's magnetic field, and micrometeorite impacts.

In the days that followed NASA's creation, officials approved the Mercury Project, a series of suborbital and orbital flights designed to put a human in space. "The chief justification for pushing Project Mercury on the present time scale," according to George Kistiakowsky, Eisenhower's science advisor, "lies in the political desire either to be the first nation to send a man into orbit, or at least to be a close second." Additionally, NASA initiated the Ranger program to photograph and map the Moon.

The Soviet Union continued to march ahead of the United States, launching the first rocket past the Moon and the first vehicle to the lunar surface. A Russian spacecraft in October of 1959 sent back the first pictures of the far side of the Moon. And on April 12, 1961, the Soviet Union placed the first human in space, cosmonaut Yuri Gagarin.

President Kennedy was in office for less than three months when Gagarin made his historic flight. As part of the budget process, NASA had pleaded for a boost in spending "to catch up to the Soviet Union in space performance," but was turned down by Kennedy, and the agency received only a modest increase in funding. Shortly after Gagarin's flight, Kennedy revisited the issue. He talked with NASA Administrator James Webb about the possibility of a lunar landing and was told it would cost more than $20 billion (about $105 billion in 1998 dollars). Days later, America suffered yet another foreign policy failure when the U.S. withdrew support from anti-Castro forces in the Bay of Pigs invasion, increasing pressure on Kennedy to take decisive action to meet the threat posed by the Soviet Union. Kennedy asked then Vice President Lyndon Johnson if there was any "space program which promises results in which [America] could win." Johnson reported back that Russia already had "a rocket capability for putting a multi-manned laboratory into space." It had "crash-landed a rocket on the Moon" and had "the booster capability of making a soft landing on the Moon with a payload of instruments." About a human trip around the Moon or to the lunar surface, Johnson told Kennedy "neither the U.S. nor the USSR has such capability at this time, so far as we know." Given sufficient resources, Johnson predicted "the United States could conceivably be first in those two accomplishments by 1966 or 1967."

Kennedy returned to the U.S. Capitol for a second State of the Union address in May of 1961 to announce his response to the Soviet Union. He warned the nation that Russia was ahead of the U.S. in space technology, and said it would likely "exploit this lead for some time to come in still more impressive successes." As a consequence, he stated we "are required to make new efforts on our own." America could not guarantee it would one day be first in space, the President explained, but "we can guarantee that any failure to make this effort will make us last." Kennedy then announced the goal for America of "landing a man on the Moon and returning him safely to the Earth." He said, "No single space project in this period will be more impressive to mankind, or more important for the long-range exploration of space."

Beyond Apollo

In the mid 1960s, President Lyndon Johnson asked NASA to identify potential missions to follow the Apollo program. The space agency produced a report that contained few specifics on possible future projects. NASA discussed in broad terms the option of extending stays on the Moon to "30 days and possibly to as much as 90 days." Other potential long-range goals included "an Earth-orbiting space station, a lunar base with roving vehicle, and a planetary spacecraft."

Eight years after Kennedy's announcement, Neil Armstrong and Buzz Aldrin planted the American flag on the Moon's surface. The public cheered the victory and took pride in the historic accomplishment. Much had changed in the world since America embarked on the Moon landing. The United States and Soviet Union had amassed huge nuclear arsenals that placed into jeopardy each country's population centers. The Cuban Missile Crisis nearly triggered a nuclear exchange, revealing the dangers of Cold War brinkmanship. China broke ranks with the Soviet Union, revealing a division within the Communist Party, and the United States was bogged down in the Vietnam War, placing huge demands on the federal budget.

President Nixon began steering a new course in American foreign policy. He replaced space competition with the policy of detente. Nixon hoped to ease international tensions by engaging U.S. adversaries in a dialog. He opened relations with China and traveled to the Soviet Union. Rather than build bigger and more powerful rockets, he sought to negotiate treaties to put limits on ICBMs. This dramatic change in foreign policy is reflected in the handshake between America and Russia aboard the Apollo-Soyuz spacecraft in 1975.

Weeks after Nixon became president he appointed a Space Task Group, led by Vice President Spiro Agnew, to make recommendations on what to do after the Apollo program. In September of 1969, the task group issued a report with a series of options for America's space program for the following two decades. The task group urged the construction of a 50-person Earth-orbiting space station, an "Earth to orbit shuttle," a lunar space station, and a permanent Moon base "capable of long stay-time on the lunar surface with 3- to 6-man teams." Assuming these programs were approved and funded, the task group envisioned the "eventual manned expedition to Mars in the 1980's."

No Longer Unique

Throughout the 1960's, funding for NASA was given special consideration. At its peak, about five percent of the federal budget was directed to the space program. President Nixon said we must "define new goals [for NASA] which make sense for the seventies." He stated that "Our approach to space must continue to be bold, but it must also be balanced." To emphasize this point, he added, "We must recognize that many critical problems here on the planet make high priority demands on our attention and our resources."

Nixon said the United States should not allow NASA to stagnate. "But – with the entire future and the entire universe before us – we should not try to do everything at once," he declared. From here on out NASA would be treated in the same fashion as other government agencies. Its budget would be evaluated the same as other departments. The era of blank check spending was over.

Nixon proposed six long-term objectives for America's space program. He said the nation should continue to explore the Moon, and move "ahead with bold exploration of the planets and the universe." On a more practical level, he said the U.S. should "reduce substantially the cost of space operations." We "should seek to extend man's capability to live and work in space," he offered, "hasten and expand the practical applications of space technology," and "encourage greater international cooperation."

NASA struggled with the Nixon Administration to attain funding for its next major leap into space. Reflective of the space agency's change in status, Congress canceled the Apollo missions 18 through 20. The Space Task Group, at a minimum, wanted Nixon to begin development of both the Space Shuttle and space station. But the combined programs were unaffordable when balanced with the demands of other programs. The space station would have to wait. Finally Nixon, in January of 1972, announced his support for a $6.2 billion shuttle program. He touted the new vehicle, saying it would give America "routine access to space by sharply reducing costs in dollars and preparation of time."

NASA Stumbles

It took a decade for NASA to develop the partially reusable Space Shuttle. In April of 1981, John Young and Robert Crippen piloted the first shuttle to orbit, putting Americans back into space. At the end of the fourth flight, NASA declared the vehicle "operational." The shuttle began ferrying commercial and government satellites, research laboratories, and DOD payloads to orbit. NASA predicted that over the first five years it would conduct 116 missions, but only 32 flights actually occurred.

The prospect of lower space transportation costs energized private industry. Congress updated NASA's mission statement to require the agency to "seek and encourage to the maximum extent possible the fullest commercial use of space." It also created the Commercial Space Centers program to help businesses learn how to use space for commercial applications.

NASA pressed government leaders to take the next step into space. President Reagan in 1984 announced his support for an orbiting space station. Congress subsequently allocated $150 million for studies to determine its design and scope. The station's size, capability, and dependence on new technologies would all greatly influence its final cost.

In January of 1986, NASA's dreams began to unravel. Challenger exploded soon after liftoff, grounding the fleet for 32 months while investigators searched for the cause of the accident and engineers upgraded design weaknesses. Space science probes configured to fit in the shuttle's payload bay had to be redesigned or were delayed by years until the shuttle could again be made flightworthy. Compounding the space agency's troubles, problems arose in the space station project. Early configurations were too expensive, prompting Congress to intervene and force more affordable designs.

NASA's increasingly bureaucratic structure and management were blamed for its growing troubles. The once proud organization seemed to have lost its vision. The can-do agency was ridiculed by critics as the gang who couldn't shoot straight. As a means to re-energize NASA, President Bush announced a long-term vision for the agency. On the 25th anniversary of the Apollo landing on the Moon, Bush pledged America to return to the lunar surface to establish a permanent base, then to venture to Mars.

Problems at NASA went from bad to worse. News articles began to question the need for NASA, which critics dubbed a white collar jobs program. In 1992, major programs managed by the space agency were running on average 70 percent over budget. The Space Shuttle repeatedly did not get off the ground due to technical difficulties. Not until the Hubble Space Telescope was launched into orbit did engineers learn it had a flawed mirror. And in 1993, the $1 billion Mars Observer was lost as it prepared for insertion into Mars orbit.

Faster, Better, Cheaper NASA

To get NASA back on its feet, President Bush replaced Administrator Richard Truly, a former astronaut, with Daniel S. Goldin, a top TRW executive with extensive managerial expertise. Goldin implemented a wide range of reforms, restructuring the agency from top to bottom. There are now about 7,000 fewer employees. For the most part – the space station being the major exception – programs are running on budget and the Space Shuttle is flying on time.

During the 1980s, NASA could afford to launch only a handful of planetary missions. Each spacecraft was running $1 billion or more. Goldin reorganized the Space Science Office, radically changing how it does business. In the place of large, complex, expensive probes, NASA began to design and build smaller, lower-cost spacecraft. Development schedules are limited, and many programs have funding caps. As an example of the new approach to space science, instead of replacing the lost Mars Observer with another large spacecraft, NASA instituted a decade long program in which it will launch ten smaller probes to the planet for the same amount of money.

Programs in the Human Space Flight Office were also reconfigured. The space station project was placed under the watchful eye of a single prime contractor. More than 100,000 pounds of hardware have been constructed and the launch of the first element is soon to commence – assuming financial problems in Russia can be overcome.

Efforts are underway to replace the shuttle with a more cost-effective vehicle. Goldin initiated the X-33 program to determine the feasibility of a single-stage-to-orbit reusable launch vehicle (RLV). The operation of the shuttle is being turned over to private industry and plans are in the pipeline to do the same with the space station after its assembly.

Despite NASA's many successes, funding for the space agency has declined every year during the Clinton Administration, reflecting an erosion of political support. Although Goldin has nursed the space agency back to health, its future remains far from certain. After the assembly of the station, the next human space flight project has yet to be determined. National policy promulgated by President Clinton only commits NASA to completing the station, at which time the orbiting laboratory "will support future decisions on the feasibility and desirability of conducting further human exploration activities." Unlike the Space Science program, which has a 25-year road map, NASA has no clear, long-term plan for human exploration. Many in the space community advocate sending an expedition to Mars, but the cost remains prohibitively expensive. (One NASA study pegs the price over ten years at $160-$320 billion.) As an alternative, others in the space community are urging a more affordable option – the establishment of a permanent base on the Moon.

A foreign policy crisis or comparable rationale does not exist as it did in the 1960s to compel government leaders to support a major boost in spending for our nation's space program. The Cold War tensions that spurred Americans to generously fund the space race are a thing of the past. So what is to be done? What should be NASA's future role and purpose? How can the agency provide adequate funding for the human spaceflight program while best advancing U.S. interests?

Redefining NASA

Forty years ago, NASA was established, in part, as a tool of foreign policy to demonstrate America's technological prowess. Mr. Goldin is reshaping NASA to respond to a new threat to our nation's technological leadership.

More and more countries are entering the high-tech marketplace and investing greater resources in science and technology. According to the National Science Foundation, "The global market for high-tech goods is growing at a faster rate than that for other manufactured goods, and economic activity in high-tech industries is driving national economic growth around the world." Between 1993-95, "high-tech industry output grew at over 8 percent per year – more than twice the rate of growth for all other manufacturing industries."

Our future ability to compete and win in the global marketplace – and sustain our high standard of living – will be dependent upon our competitiveness, which is directly related to investments in science and technology. In 1980, the United States produced 26 percent of the world's high-tech exports. Our market share has steadily declined and in 1995, the latest year for which data are available, exports by U.S. high-tech industries slipped to 19.2 percent. In second place is Japan with 11.9 percent, the United Kingdom is third with 7.2 percent, and Germany is fourth with 6.9 percent.

NASA is an important contributor to our nation's technological base. Space poses unique challenges for engineers. In outer space, power and mass are limited. Hardware must be highly reliable and function in extreme temperatures. Engineering spacecraft for science and spaceflight missions forces engineers to generate unique solutions, which are then transferred to industry for commercial applications. Space also is a new tool for research. In a microgravity environment in space, physical processes react differently allowing scientists to gather data they cannot obtain on Earth that enable them to improve terrestrial products.

Although NASA's overall budget has declined each year since 1992, its contribution to research and development has increased slightly. To help maintain U.S. leadership in advanced technology, Mr. Goldin is restructuring NASA to support long-range, high-risk research and pushing the agency to get out of the business of operating space programs, such as the shuttle fleet, which consumes billions of dollars annually and contributes little to our competitiveness.

Increasingly NASA is working in partnership with academia and industry on joint, advanced technology projects to leverage resources and ensure the relevancy of research. The X-33 program points in the direction of how NASA wants to operate in the future. Instead of government building a half-scale test vehicle, Lockheed Martin Skunk Works is responsible for the lion's share of the work with NASA providing engineering and financial support. Lockheed Martin is investing some $200 million of its own money. If all goes as planned, after flight tests in 1999, Lockheed Martin will be in a position to develop and fund the construction of a full-scale RLV, which NASA estimates could deliver payloads to space for about one-tenth as much as the Space Shuttle. The new vehicle would give the United States a competitive advantage in the launch of commercial payloads and open the door to lower-cost, RLV space transportation.

Commercial space investments now exceed those of the federal government. According to the Teal Group, some 1,700 satellites worldwide will be launched over the next decade, 70 percent of which will be commercial. The industry is expected to grow by at least 20 percent annually, generating 70,000 new high-tech jobs each year. Revenues by the year 2000 are predicted to exceed $100 billion annually.

It is no longer necessary for NASA to maintain a large government workforce to develop space technology. Private industry has matured, and can more efficiently and effectively manage many ongoing programs. NASA needs to step aside and let industry take the driver's seat. The space agency needs to assume a more supportive role in the same way it assists the aeronautics industry. NASA's goal should be to nourish industry to ensure it remains competitive by concentrating on the development of high-risk advanced technologies.

Unfortunately, there is resistance to change in parts of NASA. Many programs are cash cows that primarily benefit government workers. There are some in NASA who want to preserve the status quo and are thwarting Goldin's vision for NASA as an incubator of advanced technology to keep America competitive. The problem also extends to Congress where some members are more worried about protecting constituencies than promoting reforms in NASA. These reactions are understandable and mirror the problems the Defense Department encountered when it attempted to close surplus military bases. Only by creating an outside panel was it possible to overcome political pressures and introduce changes to restructure its military's bases.

The National Space Society recommends that a blue ribbon commission of outside experts be impaneled to assess public expectations of the space program for the new millennium, and to evaluate NASA's management of programs and to recommend ways to modernize the agency. The panel would be charged with finding additional ways to reduce costs, further integrate NASA's research centers with universities, and attract more industry support through pursuit of commercially relevant research and development.

As an example, efficiencies could be gained by tying each of NASA's research centers to state universities and boosting the involvement of industry. According to the National Science Foundation, "the academic sector is the largest performer of basic research, with expenditures totaling an estimated $16 billion in 1997." Industry is increasingly looking to academia to conduct long-term, high-risk research. "Industry officials have tapped this resource not only to realize the beneficial results of the research they sponsor," the NSF states, "but also to capitalize on opportunities to train future scientists and engineers, most of whom will one day be working in their laboratories."

NASA has tremendous talent and resources, and by integrating its facilities more closely with academia and industry, America's technological base could be strengthened. In space science, universities are now building spacecraft to explore the solar system. Johns Hopkins University Applied Physics Laboratory designed and constructed the NEAR spacecraft, which was launched in February of 1996, and early next year will rendezvous with the asteroid Eros to conduct a year-long study of its composition and structure.

Creating stronger partnerships between NASA and universities would enhance science, reduce costs, and improve education. Students would benefit by gaining real-life experience in the design and construction of advanced technology projects. Industry would benefit from having a better educated workforce to compete in the marketplace. Universities would benefit by having opportunities to work on cutting-edge technology projects, keeping them the best in the world. And the taxpayers would benefit from reduced costs and enhancements in the nation's competitiveness.

Funding Space Exploration

There are many convincing reasons to mount a human expedition to Mars. Such a historic mission would enliven America's spirit. It would allow scientists to search for evidence of fossil life. It would promote math and science education, and spur new technological developments. The primary obstacle to such an expedition is the excessive cost. Given limitations in federal funding and intense competition among government agencies for resources, it is very unlikely tens of billions of dollars in additional funding can be carved out of the federal budget for a Mars mission. There is no national crisis to warrant taking money from other government programs or to raise taxes to pay the cost of the expedition.

The politics of the 1960s were unique. Even President Kennedy resisted a lunar landing until forced to do so by an international crisis. Without a comparable situation, spending the huge amount necessary for a government funded human mission to Mars cannot be politically justified. Since the Nixon Administration, Congress has authorized only two major spaceflight programs – the development of the Space Shuttle (originally projected to cost $6.5 billion) and the International Space Station (originally projected to cost $8 billion). Until and unless a Mars mission can be accomplished at a reasonable price, it is going to be an uphill battle to gain the necessary political endorsements.

Given the sticker shock for a human expedition to Mars, a realistic strategy needs to be formulated to accommodate this barrier. NASA's options are limited. The agency can continue to slowly move forward, developing new technologies in the hope of making a human flight to Mars palatable to government officials. This go-it-alone strategy relies on government funding to succeed, and is thus a precarious approach. In the recent past, Congress and the Administration have not demonstrated any enthusiasm to boost NASA's budget, and to the contrary continue to cut spending, reducing resources available for research. Earlier this year, the Administration attempted to terminate all R&D programs related to a Mars mission. This trend could conceivably change, but is highly problematic given future demands for federal spending as baby boomers reach retirement age early in the next century.

An alternative option would be to lower the cost of human space expeditions by attracting commercial space investments and encouraging competition. Private industry is a natural ally of

NASA that can tap the deep pockets of Wall Street. Industry will be willing to invest in space enterprises once technologies are validated and it is possible to generate a profit.

An example of how this circuitous strategy might be employed to lower space transportation expenses is the development of solar space power (SSP). A recent NASA study concludes SSP could be technically and economically viable within as little as ten years. Low-cost access to space is a prerequisite to making SSP a reality. Other critical technologies include low-cost, high-efficiency solar energy conversion, modular and self-assembling systems, and robust structures resistant to debris and micrometeorite impacts.

Building a prototype SSP facility would spur private investment in space transportation. It is necessary to get costs down to hundreds of dollars per pound. The same technology to accomplish this goal could be applied to human spaceflight missions. Once power is readily available in space and transportation costs are substantially lowered, space-based manufacturing becomes a real possibility. Space tourism would no longer be science fiction.

Unfortunately, many at NASA have a jaundiced view about industry. They see space business as a way to generate money to invest in human exploration, a means to replace funding cut by Congress and the Administration. But American businesses do not work for NASA; NASA should be working for American businesses – to create new economic opportunities and new jobs.

NASA is investing minimal amounts to develop commercial space enterprises. It is spending only about $30 million annually to support commercial activities on the space station. For solar space power research, NASA had to be forced to invest a mere $12 million over three years. According to a recent report by the National Academy of Public Administration (NAPA), "If NASA wants to accelerate commercialization, it should provide ongoing funding to approximately 30-35 [Commercial Space] Centers. The public funding required for this number of Centers ranges between $75 million and $100 million a year."

We have only scratched the surface on the possibilities for space commerce. What the future may bring, no one can know. NASA needs be more aggressive in laying the groundwork for commercial space enterprises. It needs to accelerate its role as the vanguard of high-risk technology development, and further integrate its resources with academia and industry. The key to the future is not more government, but private investment. It is time for NASA to step aside and allow private industry to take the lead in space operations. By following this new course, the agency can best strengthen America's economic health and open the way for sustained human exploration of outer space.

After funding the exploration of the Western frontier, the U.S. government invested public resources to build roads and canals to promote commerce. It subsidized the construction of railroads and the national highway system. It is incumbent upon NASA to now serve a similar role. In addition to exploring space, it needs to be concerned about building commercial links to Earth orbit. This responsibility may not be as exciting as exploration, but it is necessary for America's future.

On its 40th anniversary, NASA needs to look beyond Goldin's reforms to understand the global changes that are underway, and position itself to enhance our nation's economic strength by promoting advanced technologies and supporting the development of a commercial space infrastructure.

UNITED STATES
SPACE FOUNDATION

Imperatives for National Security Space

Testimony of:

Elliot G. Pulham
Sr. Vice President
United States Space Foundation

The United States Space Foundation works to aggressively advance space endeavors, and trains teachers to bring the excitement of Space exploration and development into the classroom, in order to ensure a brighter future for all Americans. When we speak of space endeavors, we really refer to three integrated sectors that comprise the total space picture as we approach the millennium. These are commercial space, civil space and national security space. Historically, civil space has been the sector that has most captured the public's imagination, with high-profile space shuttle flights, dramatic discoveries from orbiting telescopes and deep-space probes and so on. Commercial space commands great respect and attention these days because in 1996, for the first time, commercial space accounted for more of the overall global space activity than government space for the first time ever and its lead continues to grow.

Yet no sector is more vitally important to U.S. national interests than national security space. In recent years it has become almost unfashionable to speak out on U.S. security interests, and with draw-downs in the national defense infrastructure, concerns about our nation's military preparedness in space — which can seem so esoteric or futuristic to the uninitiated — have seldom been illuminated and even less frequently championed in public.

The Space Foundation wishes to thank the committees for today's hearing. It is a modest, but important start on a public dialogue that must take place if our country's national security space infrastructure is to meet the challenges it does, and will, face.

Space is the future. To quote Apollo 13 hero Capt. James Lovell, founding chairman of the Mission HOME space public awareness campaign which is one of the many programs our Foundation is involved in: "space is the greatest bridge to peace and prosperity the world has ever known." People readily understand that statement in the context of international cooperation such as we see on the International Space Station program, or in the context of commercial telecommunication programs that are making truly global and accessible communications possible for the first time. But do they understand that the idea also applies to one of the fundamentals of our national security policy? The basic principle of peace through strength?

The military high ground for this generation and all the foreseeable generations to come is space. Such fundamental military requirements as intelligence gathering, weather forecasting, force projection, communication, command and control ✍ war fighting precepts that have been constants for thousands of years -- are today critically dependent on U.S. capabilities in national security space. Space is how our armed forces navigate, communicate, gather intelligence, and coordinate operations.

The lives of American men and women in uniform all over the globe ✍ from Bosnia to Pearl Harbor, from Riyadh to Washington, D.C. ✍ are dependent upon our military space infrastructure. We owe these men and women the best, and we are frankly not providing it to them.

The Space Foundation is not an advocate for any particular program or point solution, does not lobby for any contractor or branch of service. Whether in civil, commercial or national security space, we take a broad view of what the needs of our nation are in order to continue to reap the benefits of space. Selecting the best options for meeting those needs is, we feel, the domain and the expertise of the members and the staff of these committees.

Our current national security space needs, in our view, are quite clear:

1. First and foremost, improved access to space. This includes both launch vehicles, and launch ranges.

2. Second, providing for the defense of America■s space-based infrastructure ✍ both military and civilian.

3. Third, the need for more cutting-edge national security space research and development.

4. And fourth, a space technology master plan that allows for efficient use by government of commercial-off-the-shelf space systems and capabilities, and a more predictable, rational and stable regulatory environment for U.S. commercial space interests operating in the highly competitive international commercial space marketplace.

I'd like to address each of these four areas briefly.

Improved access to space. Our failure as a nation to adequately invest in the most basic of all national security space requirements is painfully and abundantly clear. The recent loss, within a one-month period, of a Titan, a Delta, and a Zenit launch vehicle, is nothing short of an indictment. The Titan is the heavy-lift workhorse for national security space operations. The Delta III, while on a commercial mission in this instance, is part of a family of rockets that is heavily relied on to perform both civil and national security space missions. The Zenit is a Ukrainian booster, yet relevant to this discussion because the lack of U.S. investment in launch infrastructure is the underlying problem that has driven American satellite makers to such unlikely foreign launchers.

In less than nine months, we will mark the 30th anniversary of Americans Neil Armstrong and Buzz Aldrin walking on the moon. That our country, which built the mighty Saturn V rockets that took men and machinery to the moon and back, should find itself three decades later without routine, on-demand, fail-safe access to space is both shameful and an utter failure in policy.

We as a nation are so badly in denial about this that we need a 12-step program. I would submit that our first three steps are:

1. We admit we have a problem that has become unmanageable.

2. Come to believe that a higher power — research and development — can accomplish for us what upgrades and other band-aid approaches have proven they can not.

3. Become willing, and financially commit, to developing new space launch technologies and systems.

Honorable committee members, we need to come out of denial. Our current national security launch vehicles are like Edsels boasting the latest in high-tech radial tires and 6-speaker CD stereos. Yes, they've been upgraded. Again and again and again and again. But underneath the chrome exhaust tips and shiny new paint they are 40-year-old designs. It is time to redesign and retool. Time for a better machine.

Next, we must address the slumbering Pearl Harbor of the millennium — unwillingness to accept that our space-based national security assets are sitting ducks that are vulnerable to attack. We need look no farther than the recent unexpected outage of the Galaxy satellite -- and its impact on pagers, cellular phones, automated teller machines, banks, retailers, manufacturers, entertainment and on and on — to understand how critical satellites and ground stations are to American interests. We need to start thinking about something we have not allowed ourselves to think about in the past: the need to protect our interests in space.

Defending national security assets in space — be they communication, navigation, surveillance or force projection — must become a priority mission. To the extent that the country relies on commercial and civil space systems as well, these, too, must be protected from military, terrorist and industrial threats.

Basic research and development into protective technologies, such as sensors, orbital propulsion systems, secure communication, radiation and countermeasures hardening, data integration and synthesis — the list goes on — must be expanded and funded with a sense of urgency.

And finally, U.S. space policy must be updated and rationalized to make it relevant to the global realities of the millennium. The Cold War paradigm which continues to drive federal decision-making in such areas as ITAR and technology transfer, is as out of place as bobby socks and fuzzy dice.

The current situation with the Sea Launch Company — an international commercial space consortium led by a U.S. company — is a prime example. Here we have a consortium of international allies — American, Norwegian, Ukrainian and Russian — working to adapt Ukrainian and Russian technology to help solve a commercial launch vehicle shortage of our own making. We are gaining access to their technology — not the other way around. Yet bureaucratic enforcement of an old Cold War trade regime has ground the project to a halt. Our allies and our own American companies are paying dearly for policy so antiquated it appears irrational.

We force U.S. satellite and telecommunication companies to turn offshore, whether to French Guyana, Baikonur, Sea Launch or China, because of our own unwillingness to field a launch infrastructure that can meet the demands of national security, commercial and civil space. And then, having left our own industries precious few options, we penalize them when they turn to offshore suppliers.

We need to treat our space industries with respect and encouragement. They represent the leading edge of a $120 billion-per-year segment of the economy. We need a rational, contemporary, integrated and fair approach to space policy across all federal agencies. Because economic security is a fundamental part of national security.

Space is too important to be left to chance, or ceded to our competitors. Space is what noted futurist Alvin Toffler has called the Fourth Wave of industrial evolution, and if we do not recognize this and move to position America at the forefront, the wave will simply swamp us and leave us in its wake. We must accept that national security means space security.

We must develop a new generation of space launch vehicles that assure our country has on-demand access to the high ground of space. We must aggressively redevelop our national launch ranges at Cape Canaveral and Vandenberg to assure the military, civil and commercial operations.

We must evolve our national security space forces, forces capable of fully utilizing space and space technology for all aspects of military operations. And they must be fully capable of defending our space-based assets and interests.

We must be willing to do what all other space-faring nations have done: accept that ascendancy in space is in our national interest, and merits the commitment of national resources. We can no longer expect our space transportation companies like Boeing, Lockheed Martin, Orbital Sciences, and so on, to bear the financial burden of developing our national security space transportation systems. We cannot expect them to go head to head, without meaningful government financial participation, against subsidized foreign competitors.

We are vulnerable in space. It is our duty to the American people not to be.

The United States Space Foundation urges the committees to take strong, aggressive, progressive and well-funded steps to bring our national security space programs back into the space age.

Thank you.

Chairman ROHRABACHER. And, now, because both I and Mr. Gordon would like to be as inclusive as possible in starting this dialogue, I request unanimous consent for the record that the record of this hearing remain open for 30 days, so that additional written testimony may be inserted, and, without objection, so ordered.

Well, I guess what I'll do is start now. We are expecting Pete Conrad to join us via this teleconferencing any moment. When he does we will let him have his 5-minute statement and then start the questioning. But until then, I'd like to ask a few questions and we'll proceed until Mr. Conrad joins us.

Dan, I could ask you a question, but then you might want to—Rick was little bit harsh on NASA. Maybe you would like to say a few words about that before we proceed.

Mr. GOLDIN. I would just like to accentuate the positives.

Chairman ROHRABACHER. I've heard that song before.

[Laughter.]

Mr. GOLDIN. And that is to say I'm very proud of NASA. NASA is doing, I believe, many of the things the American public wants. The Shuttle is now operated by the USA Company. It's in the act of being privatized, but this is a very difficult process to go from government-managed to private sector management.

When the private sector looked at it, they thought they would move lickety-split. It is a very complex machine. Part of the process of turning over the Shuttle for the private sector is helping them understand the complexity of the machines. There's a great awareness that comes from plans on a piece of paper and hopes and dreams to operating a system to protect human life, and I'm proud to say, their parent companies, Boeing and Lockheed Martin, are doing a superb job in taking it over, and the American taxpayers are saving money.

Just last Friday, we announced that we're turning over mission control for robotic and for our human-operated spacecraft to a private corporation in a very tough battle. The winner: American taxpayers. We are reducing our costs by $1.5 billion. So instead of costing us $5 billion over the next 10 years, it's going to cost $3.5 billion. I think this is an incredible achievement. And we've encouraged them to incorporate the private sector into it.

We are transitioning out of operations and into cutting edge R&D. That is our intent. We want to buy things from the private sector, and we started buying things from the private sector without the controls that we needed. We bought the Lewis spacecraft without the oversight that industry said they didn't need. And the contractor drove a $70 million spacecraft into the ocean. A sad lesson.

So there's some balance between walking away and giving someone a voucher to do something, and making sure it gets done right, because there's a lot to be learned, and NASA has a tremendous amount of knowledge.

Chairman ROHRABACHER. Dr. Galloway's testimony touched on that. Dr. Galloway, if you had to give NASA a grade, as you've watched it now over the years, what would you give NASA? A, B, C? Rick might give them even less than that.

Ms. GALLOWAY. I'd give them an A-plus.

Chairman ROHRABACHER. An A-plus.

[Applause.]

Mr. GOLDIN. I'd like to add——

Chairman ROHRABACHER. Go right ahead.

Mr. GOLDIN. One more element, which I think gets to the heart of this, and that is there are a number of rocket companies that are starting up now looking at low-cost rockets. I think they're going to do wonderful things. But I caution that system gimmicks will not get us to space tourism. System gimmicks will not open the space frontier. And there are some very clever things that people are doing like towing spacecraft behind an aircraft, or using helicopter blades to avoid having a turbo pump, a variety of these approaches. But in the end, it is driving technology that will cause revolutionary change. We're looking for a factor of a hundred reduction in cost, and a factor of 10,000 improvement in reliability.

And later on in the hearing, I'd be very happy to talk about these approaches we're taking towards that end. And it will involve a partnership between industry and government. And also for the record, NASA has no desire of building any government launch vehicles. We'd like the private sector to, so long as they can meet our requirements.

Chairman ROHRABACHER. Thank you very much. Jim reminded me that the other factor, of course, driving costs down would be not only technology developments, but competition.

Mr. GOLDIN. Yes, I love competition. And I'd like to see more and more and more of it. And the other factor was touched on by other members of the panel, the cost of money for risky ventures like going into space is incredibly high. We have found, in talking to the investment community, it's much less risk for them to build a chip factory, another one and much more profitable. This is a place where I believe the government is going to have to play a role. And we support government-backed loans, not for just the big companies, but for mom and pop companies. This occurs in the shipbuilding industry that came out in the hearing the other day. So it is a combination of things that will be required. And I agree, we want to see competition.

Chairman ROHRABACHER. We are lucky that Mr. Goldin provides, as I mentioned in my opening statement, a unique perspective from many of those people who headed NASA in the past in that he was very successful in the private sector, in a private sector company.

My time is about up, but let me—and we will have a second round. But, Mr. Goldin, about the financing end, I have proposed, and Speaker Gingrich agrees with the concept of restructuring our tax system so that new enterprises in space would benefit, for example. We call it the "zero gravity, zero tax" approach. Is that something that you believe would help raise money from Wall Street, and could we expect Wall Street to finance some of the things that Government has been financing?

Mr. GOLDIN. I believe it's going to take a combination of four things, and that's one of the elements. First, and I put as highest priority, partnership with industry on cutting edge R&D. We've proven that this works in the aeronautics sector. This is what the NASA aeronautics program has been about for more than half a century.

Secondly, I believe government-backed loans with appropriate requirements, such as the private sector puts in the first dollars, as being critical. And some tests for technical readiness of the product and certifications.

Third, I think government purchase of services, rather than products.

And then, an exploration of tax holidays and tax credits because tax holidays and tax credits apply to different types of things. And I think it is the entire tool kit that does it. The integration of those things, and I think we at NASA would be very happy to work with the Committee on those subjects.

Chairman ROHRABACHER. Okay. Mr. Gordon?

Mr. GORDON. Mr. Goldin, in your usual enthusiastic way, you laid out a broad agenda for NASA. Could you narrow that down over the next 10 years, what you think the top one or two priorities should be?

Mr. GOLDIN. The top development priority for NASA in working with the private sector is to solve the space transportation problem. Because unless we can get much more reliable, much lower cost transportation on a much broader basis, we're going to be forever doomed to having very expensive programs, and programs where you always worry about safety of the people involved. That, in my mind, is the number one development priority.

And I would like to make you aware of the fact that at NASA, we have told our people our priorities at NASA in order are: safety of our astronauts, number one priority; and the population around launch complexes and the employees on the ground. Number two priority is get the Space Station built. Number three priority, and number one development priority, is solve the space transportation dilemma.

We've set a goal that in 10 years, we will cut the cost of access to space by an order of magnitude and improve the reliability by a factor of 10. And in 25 years, we'll improve the reliability 10,000 times, and cut the cost by a factor of 100.

In my mind, there is no higher priority than that, and it will enable all the science, all the Earth science, all the space science, all the robotics, the technologies that will come out of that are absolutely crucial. And it is also crucial that the government maintain the role of cutting edge R&D and be supportive of the other policies we talked about, and have the industry do the actual product development because it will be the commercial sector that will determine the designs of most of these vehicles, and we just want to be another customer as a service.

Mr. GORDON. Thank you. And I want to also concur with the Chairman's accolades for the professionalism and energy that you've brought to that.

Dr. Galloway, in your statement, and I don't mean to poorly paraphrase you, but it was my sort of impression that you put forth that you think probably the foremost benefit of the space program has been that it was set up in a way to try to be for peaceful purposes, and that it has been able to maintain that. If you look back at [Alfred] Nobel, he invented TNT for—I'm sure he thought for—good purposes of building roads and bridges. But it was inevitable, I guess with mankind's primal interest in survival and aggressiveness, that it would be used for other things. And he regretted that and tried to make up for it. With the powerful tool that space can be, is there a way, and if so, how do stop those same kinds of instincts from taking us to a dominating military position there?

Ms. GALLOWAY. Well, what we did in the beginning, when we knew that this rocketry could be used for both peace and war, was to adopt policies of international cooperation because space activities are naturally international. And we preempted outer space for peaceful purposes. We got so many peaceful purposes up there that all of this billion dollar space communication industry; they want to keep it peaceful. None of the things that are prognosticated for the future could go ahead if we had a lot of disruption and conflict in outer space. So I think that the policies that we have that have maintained peace are absolutely remarkable.

Mr. GORDON. I think it is remarkable. But what happens if China, France, or the United States thinks that maybe one of the others is getting a lead on them? Maybe that they are developing some collateral types of technology that could be used for military purposes? And then they decide, well, since this other country is doing it, shouldn't we do it also? Can we afford to wait and then find out that they've done something else? I mean how do we avoid that?

Ms. GALLOWAY. Well, I think there's always a danger of outer space being disrupted. And the thing that keeps it going on a peaceful base is that everybody benefits from the way it's going now. They benefit from space communications, from meteorology, from remote sensing that can help preserve the environment of the earth. And nobody wants to destroy those things. If one nation were to destroy it, it would destroy it for its own self. It could no longer enjoy any benefits. And I think that is what keeps it going, plus all of the benefits that we have produced, many of which are quite profitable. And if they aren't profitable in the sense of making money, they're profitable in the sense of improving the functions that you are performing like space meteorology, weather reports, and things like that.

Mr. GORDON. Well, I hope you're correct.

Ms. GALLOWAY. I do think it's remarkable that when we're celebrating 40 years of NASA, and we have accomplished this worldwide, that peace is simply taken for granted. It's not even mentioned. I find that startlingly because when the Sputnik was first launched, all of Capitol Hill was absolutely struck with fear. Everybody was simply scared to death. And so we began very quickly working on these positive things that we could get going, through international cooperation and we have continued to do so.

Mr. GORDON. Thank you for your vision.

Chairman ROHRABACHER. Thank you. Thank you, Dr. Galloway, for those very fine and inspiring remarks.

Face to face with the future is what I said this was going to be all about at this hearing today. And thanks to some brand new technology, and videoconferencing technology, which is being demonstrated here today, we are actually looking the future in the face. And the future looks like Pete Conrad sitting in my office out in California.

Pete, can you hear me? We can't hear you, Pete. We're going to have to make sure—this is actually the first demonstration of this technology. And this particular type of video phone technology has been—we're going to get his voice up there any moment. And this type of two-way audio visual participation has been sought after for a long time, but videoconferencing has been expensive, it has been flawed with disjointed picture and sound, but with an investment of time, creativity, and resources, breakthroughs can be achieved. And perhaps that's what we're witnessing right now.

[Laughter.]

Chairman ROHRABACHER. Perhaps this is a major communications breakthrough, but we're going to have to hear your voice instead of just seeing your picture.

Mr. CONRAD. Do you hear it now?

Chairman ROHRABACHER. There you go. And this works just like my TV set back home.

[Laughter.]

Chairman ROHRABACHER. Well, we have with us as a witness today, Pete Conrad, a hero of both the past in space, and I would argue that he's well on his way to being a hero in the future of space as well. He's a friend. He's a constituent of mine out in California. He is here with us via this new videoconferencing technology. And he is in my office in California, and will testify. And, Pete, why don't you move right ahead with your testimony, and then we will move forward with the questions for the rest of the panel.

TESTIMONY OF CHARLES CONRAD, CHAIRMAN, UNIVERSAL SPACE LINES

Mr. CONRAD. Well, thank you very, very much, Chairman Rohrabacher, and Congressman Gordon. And thank you for inviting me to be here today.

Like all good technological programs, you do get a snag every once in a while, and as I understand it this was a typical one where the system checked fine yesterday, but somebody pulled a few plugs last night. And we are going over quite a distance, so it took them a little while to put it back together.

Chairman ROHRABACHER. That never happened to you before in any of your other endeavors, did it?

Mr. CONRAD. Oh yes it did.

[Laughter.]

Mr. CONRAD. Been there, done that, so to speak. But I'm very pleased to testify to your Subcommittee today. And I'm one of the fortunate people that has been able to fly in space, and now what I'm trying to do is get to where I can fly again in space myself. And I think that I'm probably going to have to build my own vehicle, not having served as long in the U.S. Government as Senator Glenn has, I don't think they're going to give me a ride. So if I want to get back there, I'm going to have to do it myself. And I want to do it through the commercialization of space that is now upon us.

One of the things that I would like to say is that commercial space is forecast by people that are a lot smarter than I am to be over $100 billion coming up in the next 10 years. And one of the things that we need to do is to help improve that number by making the access to space even cheaper.

Just a quick example, if you will look at the fellows that have really pushed commercial space into the forefront right now, it is the communications people. And they are doing that while it is extremely expensive for them to fly in space. Just off the top of my head, numbers, a system like Iridium, which has got 66 satellites in orbit, costs $3.5 billion or more to put up. Forty percent of those costs are in getting the system up there. One of the things that I'm working on is trying to bring that 40 percent number to a much more reasonable level. There's been talk about Reusable Launch Vehicles. We are still using expendable launch vehicles. I think that we will be continuing to use expendable launch vehicles for some period of time, as the Reusable Launch Vehicles begin to come on.

That sort of takes me to the role that everybody wants to sort out here, is: what should the government do in this business of helping us to commercialize space? One of the big things is the government shouldn't do harm to those that are trying to get there by making improper policies, or not well-thought-out means of enhancing the commercialization of space.

NASA certainly over the years has done a tremendous job in the research and development end, and the technology development end. They should certainly continue to do that in the future. If you look at the first 40 years of aviation where it went, and you look at the first 40 years of space, we haven't gone as far. And the big reason is that there is not, to date, cheap access to space.

Being an explorer—at the expense or my bill was paid by the taxpayers; and as Neil Armstrong so well said that to help all mankind—I do go back to thinking about the explorers of the early days. And people like Christopher Columbus did not really sail across the Atlantic for the good of all mankind. He sort of went across there to make a dollar. And he was funded by the Spanish government who was trying to figure out how to make a dollar. And if you think of a lot of our exploration, it has been fostered with those objectives in mind, to find some way to provide a service or do something that benefits all people, but at the same time allows you to earn a dollar to expand your business and continue to advance on all fronts.

So NASA needs, I think, to concentrate on that in the next 40 years. I think the government should look very hard at getting out of the operations end of business. And I think that we have a great future in front of us. I think we can catch up to where we should be today. I have to say that I am somewhat disappointed in how we've progressed over those 40 years, but I'm glad to see that it is really beginning to get back on track.

And with that, I wish NASA and the regulators a great 40 years of the future: opening up, I think the vistas of commercial space.

[The prepared statement and attachments of Mr. Conrad follow:]

NASA 1998:

Life Begins at Forty

Testimony Before The
Subcommittee on Space and Aviation
of the
House Committee on Science
Congressman Dana Rohrabacher, Chairman

Charles "Pete" Conrad, Jr.

Chairman and Chief Executive Officer

Universal Space Lines, Inc.
1501 Quail St., Suite 102
Newport Beach CA 92660

October 1, 1998

NASA 1998:
Life Begins at Forty

Testimony Before The
Subcommittee on Space and Aviation
of the
House Committee on Science
Congressman Dana Rohrabacher, Chairman

Charles "Pete" Conrad, Jr.
Chairman and Chief Executive Officer
Universal Space Lines, Inc.
1501 Quail St., Suite 102, Newport Beach CA 92660

October 1, 1998

Good afternoon Chairman Rohrabacher, Congressman Gordon, and other honored members of the Space and Aeronautics Subcommittee. I'd like to thank you for inviting me to speak to the Subcommittee about the future, and the role NASA can play to develop that future. Having been a long time NASA team member on Gemini, Apollo and Skylab, I rode the wave of public support and popularity the U.S. space program engendered through the 1960s and early 1970s. It was a crazy time, heady with the excitement of adventure and new worlds to explore. But I'd like to go on record as saying those days are not half as exciting as the coming age of commercial space. The American spirit has always been moved by adventure, but it is *driven* by the vision of making money.

I enjoyed the rare opportunity of being an astronaut for this great country, but the bigger legacy I hope to leave behind is a robust commercial space industry making money for America in the 21st Century. I can't speak for the entire industry, but I would like to speak for my part of it, Universal Space Lines (USL). USL is a small business just over two years old, but already with over fifty employees. Our long-term company goal is to position ourselves as the world's premier provider of affordable commercial space transportation services, including purchase and operation of both expendable and reusable launch vehicles. Our current products range from the commercial tracking and commanding of satellites, to a near term, low cost expendable launch vehicle for

53-067 99 - 6

small to medium payloads. And Mr. Goldin will be interested to hear we've begun planning for the eventual transition to reusable launch vehicles as their technology matures.

Our success will primarily be driven by the growing commercial space sector. Commercial space revenues will exceed $100 billion annually at the turn of this Century, a figure far greater than today's combined NASA and Air Force space budgets. And remember: this new millennium is only 15 months away!

As many as a thousand or more new commercial communications satellites will be placed in orbit during the next decade, extending the World Wide Web into the sky. Iridium, Globalstar, Teledesic and others are literally betting tens of billions of dollars on the opportunity to cash in on an annual trillion-dollar global communications market.

My company and others are gambling we will be a part of the emerging commercial space industry. However we should not become too sanguine about the power of the word "commercial." Both NASA and the Defense Department will also play a major role, for good or for bad, in the ultimate environment that emerges. In the years ahead my hope is that this Congress will help guide our nation to establish a free and competitive market in which all companies can participate fairly. NASA, if it so chooses, can be a major player *helping* the transition to a commercially focused profitable space industry.

As an example of how our country dealt with a similar issue from our past, I'd like to draw your attention to the early history of commercial aviation. Between the late 1940s and early 1960s, during a post war era of declining budgets, NASA (and its predecessor agency, the NACA) and the Air Force invested in a host of experimental aircraft that opened America's skies to military and commercial aviation. In particular, experimental and military jet aircraft spawned the thriving commercial aviation industry we have inherited today.

During those early pivotal years after World War II, visionary leaders in the Air Force and NACA pursued a technology policy of building and flying demonstration hardware; hardware that was built quickly and flown often. These early investments pushed aviation into a thriving, commercially focused and profitable industry. Our challenge today is to ensure the same opportunity is afforded our budding commercial space industry. Just as the success of our aviation industry hinged on the introduction of affordable and reliable aircraft, the commercial space industry can't truly take off without affordable and reliable launch vehicles.

Forty Years Hence: Through A Glass Darkly

Mr. Chairman, history is a funny thing, full of unexpected discontinuities. So before I try to look forward into the middle of the next Century, I'd like to briefly look back to the middle of this Century.

Forty years after the Wright Brothers first flew at Kill Devil Hills, B-17s and B-24s were bombing Germany, and the B-29 was in initial full scale production. In Germany, the Me-262, a jet fighter (and probably the finest airplane in the war) was also just entering initial full scale production. So, too, was the A.4 (the V-2) – an honest-to-God war rocket.

But we haven't seen the same sort of progress in the forty years since the founding of NASA in 1958. Why? In 1903, people aboard an airplane were called "aeronauts." Forty years later, they were called "passengers." Where are the passenger tickets to space available for purchase today?

A second cautionary analogy. USL is a business being run *virtually*. We depend upon the interconnectivity of the Internet. I have no idea how I would do my job without access to the information resources of the World Wide Web.

But the Web only came into existence around 1992 – just six years ago!

And we're not at all unique – scores of other businesses are also now totally dependent upon the Web's existence.

How do you predict the coming of something like the Web? It's roughly equivalent to being able to predict, in 1900, that the coming of the automobile is going to lead to the suburb, or to drive-through fast food stands…

I'm a bit reluctant, then, about trying to predict or describe what 2038 might look like. But I can describe what I'd like it to look like.

Strategic U.S. Goals In Space For the Next 40 Years

The committee has asked, "What should be the strategic goals of the U.S. in space for the next forty years?" I think that there are four overarching goals. (1) Foster a commercial space industry. (2) Explore the Solar System. (3) Settle the Solar System. (4) Explore the Universe.

For the first time, there now exists a nascent commercial launch services industry. It came slowly into existence during the last part of the 1990s, and it came into existence primarily because, for the first time, NASA didn't try to strangle this new industry in its cradle. The foremost thing a medical doctor learns is "First, do no harm." This prime principle of medicine should also become the foremost policy of the Federal Government with respect to the newborn commercial launch industry.

Exploration of the Solar System will be done by robots and by humans. In the case of robots, these missions will be primarily scientific, and could be pursued by the Government, or by academia, or both. Commercial data purchase is one method that either or both could pursue as a means to achieve their exploration goals, and at the same time save money, and again at the same time help to foster a commercial space sector.

Exploration by humans will probably be confined to the inner Solar System over the next forty years – i.e., Luna, Mars, and the small bodies (asteroids). These explorations will also be primarily scientific, certainly so in the case of Mars, but in the case of Luna and the asteroids, one can easily see economic rationales. There are thus business cases that can be made and that will be pursued.

Settlement of the Solar System may begin with Luna. There's lunar water ice at both poles, making settlements and outposts on Luna tremendously easier to accomplish than might have been otherwise. Lunar water ice, in a phrase, changes everything. One might even speak of a lunar "Cold Rush…"

The exploration of the Universe is primarily a scientific one, using space-based astronomy facilities. Such work, of course, is done to "do" science, but a lot of this science will begin to lay the groundwork for the first robotic missions to the near stars, possibly in the 22nd Century.

The Single Issue That *Must* Be Addressed

But before any of the above can be attempted, much less accomplished, there must be Cheap Access to Space. You need to be able to get to low Earth orbit ("LEO") easily, frequently, reliably, and cheaply. There is no inherent technical barrier to the creation of such a capability – "only" engineering development need occur for cheap, easy to operate, robust access to low Earth orbit to become available.

And as has been pointed out, once you're in LEO, in terms of energy, you're halfway to anywhere else in the Solar System.

Roles Of The Federal Government

The second issue the Subcommittee wished addressed is "What are the appropriate roles of the federal government in pursuing those goals?" I would argue that there are four roles for the Federal Government. The first appropriate role is to support and encourage science, both directly funding it as well as helping to encourage and underwrite its accomplishment by the private sector and academia. This also applies to exploration activity, both human and robot. The Government ought to help academia and the private sector explore, through underwriting, partnerships, tax credits, and other such mechanisms. In some rare cases, the Government itself might also mount its own explorations. These were the patterns and methods of exploration employed by Spain and England in the 1500s and 1600s, as well as by the United States in the 1800s.

The second appropriate role of the Federal Government in my opinion is to foster long-term, high-risk technology development. The Federal Government should strongly invest in next generation technology, including experimental reusable launch vehicles and military demonstration hardware

The third activity that I feel is appropriate for the Federal Government to pursue is that of the use of space for the defense of the United States.

Finally, the Federal Government has, I believe, an important, if not critical, role in the encouragement and incentivization of the growth of the nascent entrepreneurial commercial launch industry.

Short Term Policies To Accomplish These Goals

"What policies and priorities should Congress and the Administration be putting in place in the near term to begin the transition to the future?"

Here are a few of the possible options I think would go a long way in the short term for encouraging and incentivizing the growth of our emerging commercial launch industry.

NASA and the Air Force should procure all launch services *via* competitive bids that are truly open to all companies, not just the largest defense contractors. These "fly before buy" launch service contracts must not develop new launch vehicles; instead, they should be structured like the Air Mail "service" contracts of the 1930s to encourage private investment. During the next forty years NASA should transition totally out of operating space launch vehicles, or of on-orbit support infrastructure.

Space science data should be purchased by NASA in order to help to support science *and* the development of a commercial space sector. Resupply and support of the International Space Station should be provided commercially by the private sector, so as to also help support the development of a commercial space sector. The International Space Station should also be commercially operated.

In parallel, Congress can also pass legislation providing incentives to the commercial space transportation sector. One possibility is investment tax credits to incentivize the creation of launch service providers. Such credits ought to be able to be traded. Other possibilities include interest write-offs, legislated market incentives like "air-mail," and regulatory improvements. All of these incentives can help give birth to a thriving commercial launch industry modeled after today's aviation industry. The one thing we must not do is create a monopoly where a single company controls the ability to launch critical commercial and military assets into space. Guaranteeing government loans or market share for a single company would be catastrophic to the emerging commercial industry.

In the future tax credits may also be an appropriate mechanism for helping to encourage long term goals, such as Lunar missions and settlement.

A third policy thrust should be to robustly invest in the experimental technology and military demonstration hardware that supports truly low cost space launch vehicles. No technology investment is required for expendable launch vehicles, as the commercial sector is well positioned to develop such vehicles today. Instead, the government should be investing in the longer term, higher risk reusable launch vehicle technologies that promise to reduce launch costs by two orders of magnitude.

Mr. Goldin at NASA has already done a good job with his early investments in experimental vehicles, but it's just the first step. NASA's early, but under-funded plan to fly many "Future-X" experimental vehicles is an excellent blueprint for the future. In the past, Mr. Goldin has shared his vision of "blackening the sky with X-vehicles" -- not prototypes or commercial vehicles, but pure *experimental* demonstrators. If we truly want low cost launch vehicles, it will require the flight of many experimental vehicles built by many different companies.

The policy goal of flying X-vehicles for technology demonstrations should become the basic way that the government (and NASA) should approach technology development. Build 'em, fly 'em, and break 'em – both by intent and accident, this approach has led to today's thriving commercial aviation industry.

In coordination with NASA, DoD should also be investing in their own experimental vehicles and early military demonstration hardware. Either the Air Force or the Navy should develop a Military Spaceplane capability that supports global reach and the ability to defend U.S. interests "anywhere, anytime," with *dramatically* smaller force structures than exist today. Blue ribbon panel after blue ribbon panel has advocated the need for such technology investments, starting with General Moorman's Space Launch Modernization Panel in 1994. Most recently, the Defense Science Board is recommending an ongoing investment in the Space Maneuver Vehicle flight tested at Holloman AFB just last month.

Finally, while **institutional changes are not necessarily required at NASA**, the mindset must change. ***NASA should be the leading advocate of change and the transition to a primarily commercial space industry.*** Nonetheless, the real change is up to Congress. NASA, the Administration, and Congress must decide to place **funding and budget priorities** on the side of change. The Government should be investing in technology, experimental vehicles, and military hardware for the defense of the country.

2038: Free People In Free Space

The United States is at a seminal point in our transition to a commercial space industry. If we choose to encourage and incentivize the move towards a commercially based space industry we can accelerate and fundamentally enable America's move into space. We did this once before when America invested in the technology of commercial aviation, and it paid handsome dividends. Now it's time to build the same bridge to the future of commercial space.

Thank you, Mr. Chairman, for this opportunity to present USL's views. I would be pleased to answer any questions you or any other Members might have.

CHARLES "Pete" CONRAD, JR.
AEROSPACE AND PROFESSIONAL EXPERIENCE

Following graduation from Princeton University, Mr. Conrad entered the Navy and became a naval aviator. He then attended the Navy Test Pilot School at Patuxent River, Maryland, where he was assigned as a project test pilot. Conrad also served as a flight instructor and performance engineer at the Test Pilot School. After completing his tour of duty at Patuxent River, he served as Instructor Pilot in F-4H Phantoms on VF-121 and was then assigned duty in VF-96 on board the USS Ranger.

In September 1962, Mr. Conrad was selected as an astronaut by NASA. His first flight was *Gemini V*, a flight which established the space endurance record and placed the United States in the lead for man-hours in space. As commander of *Gemini XI*, Conrad helped to set a world's altitude record. He then served as commander of *Apollo XII*, the second lunar landing. On Mr. Conrad's final mission he served as commander of *Skylab I*, the first United States Space Station.

After serving 20 years in the U.S. Navy (11 years as an astronaut in the space program), Mr. Conrad retired from the Navy and accepted a position as Vice President-Operations and Chief Operating Officer of American Television and Communications Corporation (ATC). At ATC he was responsible for operating existing systems and for developing new cable television systems. In 1976 he resigned from ATC to accept the position of Vice President and consultant to McDonnell Douglas Corporation. In 1978 he became Vice President of Marketing and was responsible for all commercial and military sales for the Douglas Aircraft Company. Mr. Conrad then became Senior Vice President-Marketing in 1980. He was appointed as Senior Vice President-Marketing & Product Support in 1982, and then, in 1984, he was named Staff Vice President of International Business Development for McDonnell Douglas Corporation.

In 1990, Mr. Conrad joined the Space Systems Company of McDonnell Douglas and participated in research and development of Space Station Freedom. He then became part of the design team and subsequently Flight Manager for the Space Systems Company's Single-Stage-To-Orbit vehicle, the Delta Clipper-Experimental or DC-X. After successfully completing its series of flight test, the DC-X was modified, renamed the DC-XA, and turned over to NASA.

Mr. Conrad retired from McDonnell Douglas in 1996 to form Universal Space Lines (USL). Under the leadership of Mr. Conrad, USL has developed an innovative business strategy to develop commercial infrastructure for the growing commercial space industry. USL's long term plan is to provide launch services to space customers operating launch systems in much the same manner as an airline operates commercial aircraft. To effect deployment of this plan, USL has formed three subsidiary companies: Universal Space Network (USN), Rocket Development Company (RDC), and Universal SpaceWare (USW).

Chairman ROHRABACHER. Pete, you can see us, is that right?

Mr. CONRAD. That's correct.

Chairman ROHRABACHER. All right. Just to let you know, Dan Goldin is with us today. And earlier he made some of the same points that you were talking about. So we're going to move on with our questions now with the rest of the Committee, and it'll be back to me after a couple of the members are done.

Mr. Roemer?

[The prepared statement of Mr. Roemer follows:]

TIM ROEMER
THIRD DISTRICT, INDIANA

COMMITTEE ON EDUCATION AND THE WORKFORCE

COMMITTEE ON SCIENCE

E-MAIL: tim.roemer@mail.house.gov
INTERNET: http://www.house.gov/roemer

Congress of the United States
House of Representatives
Washington, DC 20515–1403

WASHINGTON OFFICE:
2348 RAYBURN HOUSE OFFICE BUILDING
WASHINGTON, DC 20515–1403
(202) 225-3915

DISTRICT OFFICE:
217 NORTH MAIN STREET
SOUTH BEND, IN 46601
(219) 288-3301

CALL TOLL FREE
1-800-439-3518

Statement of the Honorable Tim Roemer
Subcommittee on Space and Aeronautics
October 1, 1998

"NASA at 40"

Mr. Chairman, today marks the 40th anniversary of the establishment of the National Aeronautics and Space Administration by an Act of Congress on October 1, 1958. I am pleased that this Committee is taking the time to recognize NASA's numerous accomplishments and its future as we approach the 21st century. Americans share a sense of national pride that NASA remains the unparalleled world leader in space technology, aeronautics research and in space exploration. I salute the thousands of NASA employees who help to make this possible.

Today, our nation's commitment to expanding the development of technology and learning more about our vast universe remains as strong as ever. I believe our nation's space programs represent important investments in America's future. While I have been a strong and consistent opponent of the international space station, I continue to support NASA's mission. I strongly support its work in the vast array of space-related fields such as space science, Earth science and remote sensing, and life and microgravity science.

Of NASA's many successful missions, the Hubble Telescope and the Mars Pathfinder missions most recently captured the public's attention with fantastic new discoveries. The mobile geological studies of Mars and the images provided by Hubble are extremely innovative and educational. I would like to again commend the brilliant scientists and engineers of NASA for the successes of these missions. Additionally, NASA scientists have made dramatic discoveries in various space-related fields such as space science, Earth science and remote sensing, and life and microgravity science. NASA's Earth Sciences Enterprise and the Earth Observing System have provided, and are continuing to provide, key data on the Earth's global climate change.

In aeronautics, NASA's can be proud of its many achievements in high-speed research, advanced subsonic technology, and technologies designed to improve the safety and efficiency of our commercial airlines and the air traffic control system. NASA continues to work on new technologies to improve performance, increase safety, reduce engine noise, and assist U.S. industry to be more competitive in the world market. NASA's efforts in these areas have contributed significantly to the nation's scientific and technical knowledge, international cooperation, environmental health, and economic competitiveness. We need to do all that we can to encourage NASA to aggressively advance civil, commercial and national security space endeavors for a brighter future in space.

Today, on NASA's 40th birthday, we recognize and honor those visionaries, entrepreneurs, and leaders who have made great accomplishments in the advancement of technology through space-related endeavors.

Mr. ROEMER. Thank you, Mr. Chairman. I want to welcome Mr. Goldin to our hearing today, and congratulate him and all his employees at NASA for 40 years. I know you haven't quite served there 40 years. Some days it may seem——

Mr. GOLDIN. It feels that way.

[Laughter.]

Mr. ROEMER [continuing]. —Like that to you, Dan. But I want to congratulate you and the employees at NASA that have worked so hard over such a long period, and devoted themselves to so many successful enterprises and exploratory efforts on the part of the American people. I think a lot of us salute those efforts and, while we don't agree with them 100 percent, I personally don't agree with 100 percent of what NASA has done, I congratulate them on many of their fine, outstanding accomplishments over those 40 years. And I want to commend you for your hard work in your tenure as Director and Administrator of NASA.

I know that nobody else on this panel has probably experienced this. I was just talking to my young colleague over here from Texas to my right, but I recently turned 40. And there's both good news and bad news with that. You get a little bit older, and certainly hopefully wiser and more intelligent. You also run into some health care problems in the aging process, and you can't run as fast as your kids. And there are other problems.

One of the problems that I worry about with NASA is the Space Station. Dan, you listed it as your second priority. And we're going to have a hearing on it next week. Jack Lew, the Director of OMB, was up here about 2 months ago, and there was some question as to whether or not we could even come up with $50 million for some of the Space Station cost overruns. Just recently, NASA proposed $650 million to buy out the Russian Space Agency. Where do we come up with that money?

Mr. GOLDIN. First let me say, you read a newspaper article.

Mr. ROEMER. I hope I'm wrong. And I hope it's not that expensive.

Mr. GOLDIN. What NASA has done is we have asked for permission in a recent operating plan to the Congress to buy $60 million worth of what we consider to be very important services so we could get early research going on the Space Station, and to be able to buy storage space on board the Russian Service Module. We are doing this because A, we want it; but B, the Russians need to get that Service Module up and not hold up the rest of the schedule. That $60 million was proposed to come out of Space Station reserves that we had in 1998, and that letter is up on the Hill for its consideration.

We are in the process of planning for the Fiscal 2000 budget. And since that planning process is discussions with the Administration, that data is embargoed, and I would like to wait to comment on that until we have had full discussions with the Administration and see what the President wants to do with regards to the rest of the financing. We also want to wait and see what is happening in Russia.

But in a broad sense, I will say that our Russian partners have had a meltdown in their economy. They have incredible problems. If I were running a business, and one of my contractors was going under, I would not have a policy debate about it. I would go bail out that contractor, get the program on track, and have my lawyers do lunch with their lawyers. And this is the concept that we're undertaking because we think it is the least expensive, shortest path approach to getting this program——

Mr. ROEMER. So where do you disagree with this newspaper article? Is it that you're not proposing a bailout, or it's not going to cost $650 million?

Mr. GOLDIN. I just don't——

Mr. ROEMER. You don't want to comment on it?

Mr. GOLDIN. I feel it's inappropriate to talk about the dollar value beyond the $60 million because that is part of the Fiscal 2000 budget process. But, yes, we do believe we are going to have to expend some monies to get goods and services in return for the monies we send to the Russians. It will be necessary to do it. The exact value, I don't think is appropriate to talk about at this point in time.

Mr. ROEMER. Well, the Committee is having a hearing I believe next week on this very question, and your input before we make that decision or recommendation would certainly be helpful in the next week to make some public comments on that.

Mr. GOLDIN. I will be pleased to do that.

Mr. ROEMER. Dr. McCurdy, let me ask you. NASA has always been an agency with multiple missions. They've had science, exploration, technology development, and aeronautics has always been a very important mission for NASA. Should NASA remain this multi-mission agency in the future? And then how do we relatively keep these priorities on all these different things, if you agree with that? What should be higher than the other?

Mr. MCCURDY. I think NASA will be a multi-mission agency, in particular in two areas. I think it will continue to supervise the construction of large structures in space. We've heard a lot about improving space transportation, but there are two visions that are represented on that Colliers Magazine cover that I put up there. The cover represents space transportation: easy, cheap access to space.

But the lead article in that magazine was Wernher von Braun's "Crossing the Last Frontier," which was his proposal for the development in 1952 of an orbiting Space Station, 1,000 miles above the Earth. He didn't know about the Van Allen belts at that time. But that was the lead article at that time.

So we have two unfulfilled visions: one is the development of space transportation; the other is learning how to construct—assemble—large structures in space. What we learn from the Space Station, we're going to apply to lunar bases and other structures throughout the solar system. So there's a tremendous technology development focus that has to be accomplished through the Space Station.

Now in order for NASA to do that, it has to have multiple capabilities. It has to be able to encourage private investment in space until we have another agency that does that. There may be another agency that doesn't look at all like NASA, like a financing agency. NASA is going to remain, at least for the next decade, the primary conduit through which government funds go to private industries that build these structures in space. And it's going to have to learn how to do that effectively without trying to micro-manage those industries.

When NASA got started, Wernher von Braun's rocketeers built the first Saturn rocket in-house at Huntsville to learn how to do it. And then they did what they call penetrating the contractors, making sure that the contractors built it exactly to specifications. And that was because the contractors didn't know what they were doing. The rocket talent in the United States was focused at the Huntsville center. But now that rocket talent and other talents are widely dispersed. So NASA is going to have to learn how to monitor contracts until another agency comes along to do that. Another agency may take that over in 10 years.

And, finally, NASA has to do what it does best, which is research, development, discovery, exploration. And that model, whether it's in business or whether it's in government, is the model that I mentioned before. Everybody knows how to do it. We just have to make sure that NASA is organized to accomplish that.

I'd make one recommendation if I could. I know nobody wants to hear this, but maybe NASA ought to think about trying to segregate these missions by Centers. It's so hard to do research and development and not penetrate contractors. At the same you might want to revisit the purpose of the Centers at some point to find out if there's one Center, an existing Center, or even a new Center that would be strictly in the business of overseeing the construction of large structures in space that wasn't encumbered by having to also do research and development and have people that really needed that hands-on ability at the same location.

Chairman ROHRABACHER. Thank you very much for those suggestions. And Ms. Jackson Lee? We will have two rounds of questions. Pete Conrad will be sticking with us, and he can hear your questions from your microphones.

Ms. JACKSON LEE. Thank you very much, Mr. Chairman, and let me thank Mr. Gordon, as well, for having this hearing. And it is a celebratory time. Let me say that there are many things I agree with my good colleague from Indiana, many things I do not. But certainly let me go on record for his youthful assessment of his colleague from Texas. And 40 is still young.

[Laughter.]

Ms. JACKSON LEE. And I want to congratulate NASA. I want to offer to you the congratulatory remarks from my constituents in Houston. I was speaking before the community that was born from Johnson Space Center, and Clear Lake, the Chamber of Commerce. I want to bring greetings from them.

I'd like to make several comments, and in the midst, ask the Chairman to put my opening statement in the record. I ask unanimous consent?

Chairman ROHRABACHER. Certainly, certainly. Without objection.

[The prepared statement of Ms. Jackson Lee follows:]

SHEILA JACKSON LEE
18TH DISTRICT, TEXAS

COMMITTEES:
COMMITTEE ON THE JUDICIARY
SUBCOMMITTEE ON CRIME
SUBCOMMITTEE ON COMMERCIAL AND ADMINISTRATIVE LAW

COMMITTEE ON SCIENCE
SUBCOMMITTEE ON SPACE AND AERONAUTICS
SUBCOMMITTEE ON BASIC RESEARCH

Congress of the United States
House of Representatives
Washington, DC 20515

WASHINGTON OFFICE:
410 CANNON HOUSE OFFICE BUILDING
WASHINGTON, DC 20515
202-225-3816

DISTRICT OFFICE:
1919 SMITH STREET, SUITE 1180
THE GEORGE "MICKEY" LELAND FEDERAL BUILDING
HOUSTON, TX 77002
713-655-0050

Statement
Congresswoman Sheila Jackson Lee
Subcommittee on Space and Aeronautics
"NASA at 40"
October 1, 1998

I would like to thank Chairman Rohrabacher and Ranking Member Gordon for holding this commemorative hearing on the occasion of the 40th Anniversary of the National Aeronautics and Space Administration.

NASA's vaunted history started on this day, 40 years ago, as a response to the launch of a Soviet-made satellite called "Sputnik". Although the Soviets had quite a head-start on us back then, it was not soon afterwards that it was the United States that was leading humankind into the uncharted territories of space.

NASA's earliest task was to catch us up to the U.S.S.R. That meant we needed to launch our own set of satellites with the same, or better capabilities than our competitors. The result was Explorer I, which was the first satellite to discover the Van Allen radiation belts. The following year and a half brought other significant breakthroughs through the use of the Explorer and Vanguard Programs. Those breakthroughs involved the collection of radiation and micrometeriod samples, and included the discovery that the Earth was, indeed, pear- shaped and not round.

NASA's next goal was to successfully put a human into orbit, which spurned on the creation of "Project Mercury". That effort culminated into a 15 minute, suborbital flight in May of 1961 that carried Alan B. Shepard, Jr. into annals of history as the first American in space. Just a short time later, we were able to successfully launch the first American into orbit, now-Senator John H. Glenn, Jr. It was a short flight by today's standards, totalling in at over 4 hours, but that was a tremendous improvement over the Shepard flight just less than a year earlier.

The next challenge for NASA was posed by President Kennedy, who urged the agency to "land a man on the moon and return him safely to Earth." To make the challenge even more daunting, Kennedy asked that the task be accomplished within the decade. NASA fulfilled that mandate by putting Neil Armstrong and Buzz Aldrin on the moon in 1969, almost two years earlier than Kennedy's deadline. It was a marvelous achievement that captured the imaginations of all Americans -- causing many children to change the answer to the familiar question "What do you want to be when you grow up?" from "a policeman" or "a fireman" to "an astronaut" almost overnight.

With the increasing demands of the economy and a war Southeast Asia brewing, NASA re-focused its energies to using space technology to improve the quality of life here on Earth. The Agency began to look into "remote sensing" technologies, and converted Apollo hardware into "Skylab", our first laboratory in space. These projects soon began to pay dividends, as experiments and observations by both programs told us more about our planet, and ourselves, than we ever imagined.

Although NASA was able to produce great results, the budget constraints placed upon the Agency forced them to look to new ways to explore space. The answer in the 1980s was the Space Shuttle, the first almost completely-reusable spacecraft. We all remember the elation of the first shuttle launch well. It was a poignant moment marking our triumphant return to space after too long an absence.

However, the dangers and risks that space also presents manifested themselves with the tragic explosion of the Space Shuttle Challenger in 1986. It was a moment that shook NASA to its very foundations, putting a damper on subsequent shuttle missions and reminding all of us that success should never come at the cost of safety.

Having already met many of the challenges of space, NASA next set its sights on its most difficult task yet - dealing with the demands of an ever-shrinking budget. NASA responded by changing its whole paradigm, and defied its critics by making "better, cheaper, faster" a reality, instead of just an empty catch-phrase. Even in recent weeks, we have seen NASA continue that strategy, by outsourcing their administrative tasks to companies like Lockheed Martin, and saving the taxpayers of this country over $1.4 billion over the next decade.

There are many people who would take away from NASA's successes by pointing to recent developments in the completion of the International Space Station. I support the space station, and urge the patience of the Members of this Committee, and the general public. We must remember that we are not the only participants in this project, nor could we be. We need the expertise of the Russians and our other international partners to build this station in order to maximize its benefits. Furthermore, when urging speed, we must remember the lesson of the Challenger. Safety should never be compromised for the sake of expediency.

As we move into the future, I see many more successes for NASA. We are just a few short months from the launch of the first module of the International Space Station. The Hubble Telescope is still bringing us pictures that we never dreamed were possible, and Future-X and X-33 are right around the corner. In all estimations, the future of NASA looks to be as bright as its past.

I thank all of you for coming, and look forward to hearing from all of you.

Ms. JACKSON LEE. It is important to note some of the changes that have been made. I delight in competition. And I know that the launching of the Russian Sputnik launched our commitment to speedily move into space, and I think that we have handled ourselves very well. But we've come a long way.

And as the only woman on this panel right at this time, let me take the special privilege of applauding the Chairman and the Ranking Member for having Dr. Galloway here. Even though I know that her beginning in this arena certainly had nothing to do with her gender, but I will say to her that I have appreciated her work, her leadership, and also some of the premises upon which she has suggested that we follow. And certainly the issues dealing with peace. I think that we have been most successful in peacefully utilizing space collaboratively with our fellow partners in this.

Just as an anecdote, Dr. Galloway, I just flew in from Texas, and I did all that I could to get here for this hearing. But late last night, I was trying to turn the light out on my 13-year-old who doesn't seem to know time. And he hollered downstairs and said, "I'm on the phone talking about my science exam tomorrow. I'm talking with the smartest person in the school. She's the smartest person in school, and we're studying science together." So I can applaud your leadership because the person he was talking to was a girl, and my son was trying to get her insight on how he could pass the science exam. I was glad for that, and let the light stay on a little bit longer.

But we have come a long way and I would simply ask you again about this question of the peaceful use of space, and as you look into the future, how can we ensure that happens?

And, Mr. Goldin, let me thank you for your work. I won't have a question for you. I will say that I'll follow the inquiry of, again, my colleague from Indiana. My interest in the Russian payment and a briefing on that, I'd appreciate it. And I know that you'll be working on that for us. And likewise I'd like a briefing on—even though a very good proposal [the recently amended]—Lockheed contract. I'd be interested in that as well.

But I'd like to just go to Dr. Galloway to ask about how we can ensure the continued peaceful use of space. Dr. Galloway?

Ms. GALLOWAY. Well, I think it will be ensured because of the benefits. The international space communication, for example, is a global industry. It's about $77 billion a year. It employs 800,000 employees world-wide. And it's not only the industry itself, but it has services so that it spreads out for other industries. And nobody wants to destroy this. Everybody benefits from it. Everybody benefits from the space weather reports that saves lives and property. And that is one of the biggest things that you have going is that we preempted space, the whole area of space, for these peaceful uses and they are very valuable. They have all kinds of values for education and exploration and increase of knowledge. And any country that tried to destroy it would be committing suicide itself. They would no longer have any of the benefits. So it is something that is sort of self-enforcing.

We have a treaty on principles guiding States in the conduct of their space activities. There are 93 members—93 nations are members of it. And they have given up the idea of sovereignty because that was a potential for conflict in outer space.

And then we have the United Nations working on this all the time, in connection with the Committee on the Peaceful Uses of Outer Space, so that there is a connection between the scientific community and the legal community to have international law in outer space.

Ms. JACKSON LEE. Thank you very much, Dr. Galloway. And I would simply want to congratulate John Glenn in his launch on October 29th that will help us in understanding aging, and, of course, pay tribute to President Kennedy who challenged us to land a man on the moon and return him safely to Earth. This is a celebration, so let me just simply end, Mr. Chairman, by saying congratulations to all.

Chairman ROHRABACHER. Thank you, Ms. Jackson Lee. We have time for one more round of questions. And I will just proceed. And I'm going to go straight to Pete Conrad who is sitting out there in my office in Huntington Beach. I will not ask you how high the surf is out there today, but I want you to know that my office in California does look out over the ocean.

Pete, we were talking today about how best to inspire the private sector to become involved or give them incentives to become involved and meet their needs. Dan was mentioning earlier changes in the tax law and also the possibility of loans to entrepreneurs who are going to be doing space activities. You, yourself, are involved in some entrepreneurial efforts. You might want to enlighten those of us on the Committee about some of those. But the main question I have for you is would it be better to go the route of the government providing loans to companies like your own, which you're trying to start up and get going? Or would it be better for the government to provide a change in the tax system which provided you some type of an incentive?

Mr. CONRAD. Well, I'm not certainly an expert in any of those areas, although I am trying to deal with them now in the commercial world. Certainly if the government gets involved in doing government loans, as a little fellow I get very leery having worked for a big fellow, that we make sure that if anything like that happens that the little fellow can compete on equal grounds with the big fellow to get those loans. I personally am not in favor of that. I think for the commercial industry to grow and big strong, it should not—and I look at the loan as a subsidy—it really shouldn't happen in my opinion. I think that to cut your teeth in the commercial world and really grow, part of it is convincing people to invest their money in you and the job that you are doing.

Chairman ROHRABACHER. But a tax incentive or just providing an overall tax program would not be a subsidy.

Mr. CONRAD. Tax incentives I think are very beneficial, but I am not in favor of this loan program. How do you know that the loan is being fairly meted out, who is the [loan] committee, whatever? You are in effect doing this when you go to Wall Street or to the other places to obtain financing in the commercial world, and basically you're, however you want to look at it, equity or debt financing or whatever, you've got to convince these people that you've got a solid thing.

Chairman ROHRABACHER. Pete, you might tell the Committee, inform them what you're trying to do in the private sector in terms of control systems for private satellites or for satellites?

Mr. CONRAD. Well, we are doing two things. I have two companies. One of them is called Universal Space Network. And 3 years ago, I woke up with the idea that if I was going to be involved in putting things up and bringing them back down, the commercial world needs somebody to tend their satellites when they're on-orbit because that's not the core business of the satellite. And so I found out that I didn't have a unique idea, but it turned out that although some people had thought of it, they hadn't really gone about doing anything about it. So we did. And we did on private money construct a now worldwide tracking network that's commercially available to anybody that wants to use it. And we wrote the software that makes it run autonomously and do the scheduling and processing and so forth autonomously. And we have that now up and operative. We have also signed agreements with some foreign entities that have tracking stations, and we are in the process of tying them into our network.

So we're online. And, as a matter of fact, I believe our first ad appeared today or will appear Monday in the newspapers. The Space News is advertising that service as being commercially available right now today as we speak.

Chairman ROHRABACHER. Let me note for members of the Committee that here we have a service, I mean no one would have ever thought of this 30 or 40 years ago having a private company established, making a profit in managing satellites for different countries and different private entities. And I think that this is something that on our 40th anniversary of NASA, it sort of shows that NASA was a pioneer, and now we're looking forward to another 40 years. It's going to be a whole new world opening up there.

I'd like to turn to Mr. Goldin now, and talk about the new world. And that was the bell on me, so I'll just ask one question and then we move on. Dan, do you think mankind, or humankind I should say, is going to be dependent on rockets? NASA seems to have its history tied in the past to rocket transportation into space. Do you predict that, when we celebrate 40 years from now, that NASA is going to be, and that mankind, again, pardon me, humankind's efforts are going to be, tied to the rocket, or will there be other forms of transportation and propulsion?

Mr. GOLDIN. Well, I believe that we will not necessarily depend on the combustion process taking 100 percent of our oxygen with us up into space that we've been doing for 40 years. We are now working on mag-lev launches. We're working on pulse detonation engines. For inter-planetary travel, inter-stellar travel, we're even looking at anti-matter. We're building containers that one could put anti-matter in. We're looking at fusion propulsion, and a whole broad range within that spectrum.

And on October 25th of this month, we hope to be launching the first-ever electrically propelled vehicle that will go into solar space. And if this vehicle is successful, if this technology is successful, we have opened up a whole new realm for traveling to the planetary bodies in our solar system.

And the other reason we're interested in this technology, I alluded to in my opening statement, the concept of transportation nodes which gets to the infrastructure that you were talking about. We think about transportation as leaving Earth, going somewhere, and coming back to Earth. I think what we will see in the next 40 years are a set of transportation nodes. One of them might be logically be put at the solar L–1 point, which is about a million miles from Earth. You're completely out of the Earth's gravitational well, and then you could start thinking about transportation systems that don't have to fight the Earth's gravitational force when you go orbit to orbit, and you almost fall to Mars. It's about 10 percent of the energy. Ninety percent of the energy to get to L–1, 10 percent of the energy to get to Mars.

So some of these electrically-propelled vehicles, electrically-propelled concepts could be a cargo vehicle. The Shuttle has astronauts and the Shuttle has cargo, which is one of the reasons it's so expensive. What we think we will see in the future is a separation out of cargo and people. There will be a small taxicab for the astronauts, and then, if you will, we could take a slow boat from low Earth orbit to these transportation nodes using perhaps hundred kilowatt electric generators which in turn could then be applied to generating power when we get to our destination.

So these are the technologies that we're working on, and we encourage people to get low cost means to get to space for the initial next commercial wave, which will be based on making lower-cost combustion approaches work, but in the long run we're talking about these way-out technologies that I think will pay off in the next decade or two.

Chairman ROHRABACHER. In my first conversation with Mr. Goldin when he took over his job, I remember the very first thing we spoke about is setting a priority and bringing down the cost of getting into space, and this is very exciting.

Mr. Gordon?

Mr. GORDON. Thank you. Dr. McCurdy, you had mentioned earlier in your testimony, concerning budgeting and planning, that it would be better to have a lump-sum budget and if you go over 15 percent, you pull the plug. Now oftentimes in science it's sort of the unexpected, maybe, discovery that is the benefit, more so than what you had planned for. I mean, how often do you reevaluate on your 15 percent? Are you suggesting that if it's a 10-year program, you decide, this will be our lump-sum budget for 10 years, and that we'll have benchmarks every year? And then if we come up with 15 percent, whether you are in year two or year seven, you pull it.

Mr. MCCURDY. It's always going to be discretionary; it wouldn't be a blind formula. But in general, I am in favor of the concept. Project Apollo was accurately estimated, the cost of that. I would like to see a system that obliges managers, engineers within NASA to accurately estimate costs. And second, I'd like a system that encourages them to invest as much as they can in hardware and as little as they have to in overhead.

One of the great problems with the Space Station program, which originally was estimated at $8 billion in 1984 dollars, was that NASA basically had to stand still for 10 years and spend money on overhead without producing hardware.

So the funds were still being spent, but the work wasn't being done because NASA and its contractors were in a continual state of redesign.

So I'd like to see something that encourages people to invest in hardware. And I remember talking to the people who designed the original Space Station concept, people like John Hodge and Phil Culbertson. And they were convinced—maybe not $8 billion, maybe $10 or $12 billion—they were convinced that they could build the Space Station for that amount of money.

Mr. GORDON. So would you pull the plug then?

Mr. MCCURDY. On what happened with the Space Station?

Mr. GORDON. The Space Station is obviously more than 15 percent over, so with your methodology here, do you pull the plug?

Mr. MCCURDY. If you follow my methodology from the beginning, I'd pull the plug. That is, if you had said to NASA, give us an accurate cost estimate of the Space Station, an honest one, and we'll allow you some latitude for overruns, 15, 20 percent. It's new technology; there are obviously going to be some unexpected problems. You say you are going to develop a Space Station for between $8 and $12 billion in 1984 dollars and you are going to do it within 10 years, and if you can't do it, yes, I would pull the plug. But from the beginning——

Mr. GORDON. So did those people not tell President Reagan the truth?

Mr. MCCURDY. I think they honestly believed that they could build a Space Station for what they estimated the cost. I wouldn't hold to the $8 billion because I know there was some—pardon?

Mr. GORDON. So if they were telling the truth, then the plug should be pulled under your methodology.

Mr. MCCURDY. The plug would be pulled if they were given a lump-sum appropriation or line of credit. They weren't given a lump-sum appropriation or line of credit. They had an annual appropriation and part of the difficulty that NASA had in meeting its cost goals were that it was operating under an annual appropriation.

Mr. GORDON. We've seen the recent hedge fund that is being bailed out under the idea of "too big to fail," whether it's banks or various things that cause such disruption. Do we have a too-big-to-fail situation, where other countries, lot of dollars have been spent, or if you're 15 percent over, you pull the plug?

Or do you distinguish the small projects from the major signature projects?

Mr. MCCURDY. Well, I will agree with you, it works better for small projects. But let's take what we're learning from small projects in the faster, better, cheaper area and try and transfer that as much as possible to large structures.

Mr. GORDON. So then let's take it out of hypothetical and put it into the Space Station where we are today. What should we do?

Mr. MCCURDY. Well, where you are today——

Mr. GORDON. Well, we can't be where we were yesterday. So we're where we are today. So what would be your methodology?

Mr. MCCURDY. Most of the development work has been done on the Space Station, so it's in a different phase. It's moving toward the operational phase, which is very different. But let's say you take something like a lunar base or a second Space Station.

Mr. GORDON. So you don't set a budget then for the rest of the Space Station?

Mr. MCCURDY. I think you do, but I think what you're going to have to do with the rest of the Space Station is you're entering the operational area, and then work with annual appropriations or with government contracts that tell the contractor, if a contractor is eventually going to run the Space Station, what the amount of funds are that are available for that. So there has to be some funding certainty. I think you are too far in to——

Mr. GORDON. We don't expect it to go down any; it goes up. So it's not like they're being tricked. The rug isn't being pulled. It only goes up. So what makes them nervous?

Mr. MCCURDY. On current Space Station.

Mr. GORDON. Yes.

Mr. MCCURDY. I think if you don't start with it, it's really hard to introduce that notion on a project that's already underway. Take Mars Pathfinder as an example. You can ask them the question. I think they told the people in Mars Pathfinder, you've got $250 million?

Mr. GOLDIN. Two hundred seventy four.

Mr. MCCURDY. Thank you, $274 million to run this program. That's one tenth of what it cost to send those Viking landers to Mars 20 years ago.

Mr. GORDON. So little banks can fail, but big banks can't?

Mr. MCCURDY. I would extend it to large projects. I would start with the small and see what you can do with the small. In effect, NASA is doing that with its small projects by saying this is how much you got, we'll pretty much guarantee you are going to get it. This is how you are going to receive it over the years.

You've got to keep the annual appropriations process intact, but the people who at Pasadena knew how much they had. And I think if they got into trouble and started to run 20, 40, 50, 100 cost overruns like, for example, happened on the biology package on Viking—you can ask him—the administrator says he would have pulled the plug.

Mr. GORDON. Thank you.

Chairman ROHRABACHER. Well, you know that if Mr. Roemer was here, that we'd be discussing Space Station.

Mr. ROEMER. I'm not here, Mr. Chairman?

Chairman ROHRABACHER. Go right ahead.

Mr. ROEMER. Thank you, Mr. Chairman. I think I'm here, I hope I'm here.

Mr. Goldin, let me come back to the Space Station and follow up on a couple of Mr. Gordon's comments here and I'm sure we'll be interested in your comments next week on this, as well.

A meltdown or implosion in Russia. Those are the kinds of words we used 2 or 3 years ago to warn NASA about the potential problems. NASA told us at the time that they were working on a contingency plan to address this situation directly to potentially get the Russians out of the critical path.

Many of us offered amendments to get Russia out of the critical path. Those were not agreed to by NASA or the Committee. We now find ourselves in a very, very difficult situation. Russia is imploding. They are not delegating the appropriate dollars to the Duma, to the Russian space agency. Whether it's $650 million that has been floated in the paper, whether it's more than that, we're not sure.

Do we have this plan from NASA now on a contingency plan as to what to do? Can you share that with the Committee?

Mr. GOLDIN. We had a contingency plan. We've been working on a contingency plan. One of the elements was to not have to depend on the Soyuz vehicle. We have put into the budget and are working on a crew return vehicle. We did listen to the Congress.

Two years ago, we started an Interim Control Module, which was a propulsion module which would protect us in case there would be some delays in the Russian space program. To go have a 100 percent contingency would have cost well in excess of the amount of money we will have to pay to help bring the Russians along on this particular aspect, in addition to which——

Mr. ROEMER. What is that estimate? I'm curious, because you wouldn't answer the $650 million question before. What is the estimate then to bring them along?

Mr. GOLDIN. I will talk in general terms. I would be happy to answer your questions at the hearing on the Space Station. We are having discussions with the Administration, and I think it's inappropriate that we haven't had an agreement on the budget beyond 1999 to throw out numbers that will then be in the record and then we'll come back and talk about them again.

I will again say that I believe we have taken steps. Could we have moved faster? Yes, we could have, but we would have spent money on things that we might not have had to spend. You can't have a 100 percent contingency. In addition to which, we want to encourage the Russians to work with us.

This is another issue that is a broader issue. The Russians have one real terrific thing going now, their space program. There are very few other good things going on, on a government to government basis with the Russians. We have tried to be in some way sensitive to the issues with the Russians. I do believe we have a contingency plan. Yes, it could have been more robust.

Mr. ROEMER. Is it something you can share with the Committee, because I've never seen it. Is that on paper that you can share with this Committee, so that we can make these decisions.

Mr. GOLDIN. I could tell you exactly what we are doing, yes.

Mr. ROEMER. So it's on paper and you will get it delivered to the Committee.

Mr. GOLDIN. Well, it will be in my testimony.

Mr. ROEMER. So if we'd asked you 2 or 3 years ago to get this plan put together——

Mr. GOLDIN. What would you like? Would you like me to come over and give you a briefing?

Mr. ROEMER. I just never have seen it. I'm not trying to be facetious or difficult with you right now. I've just never seen this contingency plan. I'm not sure anybody on the Committee has.

Mr. GOLDIN. Well, I believe I have testified a number of times. I will go back through it. We are developing a crew return vehicle for six to seven astronauts. We have built the Interim Control Module. We are negotiating with our contractor to get a propulsion module. We are upgrading the Shuttle so that it can do orbit-raising. And I believe those are the core elements of the plan to which I have testified to.

And as part of that contingency plan, because it would be much less expensive to have the Russians launch the Service Module in 1999 than to have us replicate that Service Module and launch it 2, 3 years later, we are proposing that we provide appropriate funds to the Russians so that Service Module can be launched.

Mr. ROEMER. So with all that, you estimate all those things, you've got to have a cost in mind as to what that will cost the American taxpayer to bail out the Russian space vehicle.

Mr. GOLDIN. There are NASA estimates on that and I would prefer to deal with it at the hearing which is scheduled next week by Chairman Sensenbrenner.

Mr. ROEMER. Okay, fair enough. Well, let me give you an easier question, then, to deal with than those cost figures that we keep on seeing going up and up.

NASA has recently been accused by Speaker Gingrich of "making space as boring as possible." Do you agree or disagree with that statement?

Mr. GOLDIN. I think that Speaker Gingrich deeply loves the space program and I think that on any given day, any human being could say some things—I don't know the total context with which that was presented. I don't think his conviction to supporting the space program is any less today than it was that day or the day before that statement was made.

Mr. ROEMER. Well, I disagree with that statement. I think that NASA has done a wonderful job, for instance, on the Mars Pathfinder, where I had constituents glued to the television set for weeks and months at a time trying to get new information as to what was coming back from Mars, what data, what pictures, what success.

And I think a lot of things that NASA does so successfully, Galileo, Mars Pathfinder, what we have learned from the Hubble telescope, and what great knowledge and information is coming back on those fronts, sometimes is clouded by the problems in Space Station.

But nonetheless, they are great successes and exciting successes that the American people are not only engaged in, but are excited about. And I hope that we can find better ways of getting that information out to the American people, so that they can be proud of NASA, they can be involved in these kinds of efforts in the future, and that we can cut through some of the haze and the clouds surrounding some of these efforts that don't get, quite frankly, some of the publicity they deserve for their success.

So I would disagree strongly with that statement, although again, I come back to NASA is doing about 90 percent of everything absolutely correctly, something we can be very proud of. Where you and I just disagree is on the Space Station, and maybe Dr. McCurdy's statements on that will be helpful to us next week.

I thank the Chairman.

Mr. GOLDIN. Could I respond?

Mr. ROEMER. Absolutely.

Chairman ROHRABACHER. Go right ahead.

Mr. GOLDIN. I am not only proud of NASA for what they are doing on the other activities you talked about, I am extremely proud of what they are doing on the Space Station. It's a very complex world we live in, and I believe when it's done, this Nation is going to be incredibly proud of it. And on top of it, this Nation is going to be proud of the leadership we are providing to the international community.

Our partners from Europe and Japan and Canada have come to us and asked for U.S. leadership to help us through this very difficult time, and we'll get through it. It's going to be difficult; it's going to be tough.

But as my distinguished colleague has talked about the infrastructure in space, we are not going to go beyond Earth orbit unless we can figure out how to take big complicated things and put them together. And in the middle of the process, it looks messy, but I'm proud of the team and I'm proud to be representing them.

Mr. ROEMER. Mr. Goldin, I wouldn't disagree with a thing you said there. I am proud of NASA with not only Pathfinder, but what they could do with the Space Station. My questions come at what cost to NASA, what cost to a flat budget, what cost at cannibalizing other programs, what cost to our international partnerships, and we are in a very difficult situation with all those questions right now.

Mr. GOLDIN. Well, I would be pleased to take each one question by question at the next hearing, because we have not cannibalized NASA. We are launching a spacecraft into interplanetary space every 11 weeks, and from 1983 to 1992, we had two interplanetary spacecraft. And this is tough stuff, and NASA is about tough stuff. So I would be pleased to delve into this deeper at the next hearing.

Mr. ROEMER. I'm looking forward to it.

Chairman ROHRABACHER. Thank you very much. Now, just to remind you, Mr. Roemer, that Dr. Galloway gave NASA an A plus, even considering the Space Station, so I wonder what it would be like without it.

We have been joined by Mr. Pickering from Mississippi, and you may proceed.

Mr. PICKERING. Thank you, Mr. Chairman. It is probably a good time to follow up on the exchange that we just had.

But first, let me commend Mr. Goldin on his leadership, not only as we have tried to work through the Space Station, but also in issues like remote sensing, which has received tremendous attention from you, and the benefits and applications that we are seeing as the Stennis Space Center tries to develop the commercial applications and contribute to our research community. The Chairman has also provided tremendous leadership in that area.

But my question goes to if we want to continue seeing the type of commercialization and the benefits that we are seeing today and the tremendous potential that we have in that area, a concern has been raised, or possible speculation, to make up on the shortfall on the Space Station that the data buy in the Earth Sciences portion of the budget could be delayed or reduced. Would you care to comment on that, if you would?

Mr. GOLDIN. I would be pleased to. The Administration and the Congress have put a firewall between Earth Science, Space Science, and the Space Station, and we concur with that. We will not be transferring dollars out of Earth Science to pay for the Space Station.

Mr. PICKERING. So the $50 million planned data buy is going forward on schedule at the same level, and there is no threat to it.

Mr. GOLDIN. I believe so, but it may not be because of the Space Station. I will have to go check. One moment. I believe it's going through. If I'm wrong, I will get back to you.

Mr. PICKERING. Thank you very much. One other question. This is also in the same context. I would imagine that you are experiencing tremendous concerns, questions, rumors, speculation on how does this affect all the other programs' timetables and different things. One that is a possible relationship to the situation that we now see in Russia, and it's probably an issue that you may not know about, but would like your Agency just to formally communicate with me concerning.

The T–38 ejection seat contract is scheduled to be determined, I believe, October 15th of this year. It has already been extended or delayed 90 days from the original time. It was to be determined July 15th, but there is some concern that this may be extended again in order to allow a Russian company to compete. I hope that would not be done in a way that would put American companies at a disadvantage or be inconsistent with the procurement and competitive procedures. Are you familiar with the T–38 ejection seat contract?

Mr. GOLDIN. It's in source evaluation now, and since I am not a source selection official, I don't know the details. But I will be happy to have our legal counsel follow up on your question and to get back to your office with a clear answer.

Mr. PICKERING. I hope it goes forward as scheduled October 15th. Mr. Chairman.

Chairman ROHRABACHER. Well, you didn't use up all of your time, so maybe we'll have Rick, who is sitting there biting his lip. Do you have a last comment there, Rick? You're the real rebel on the panel today.

Mr. TUMLINSON. Win friends and influence people.

Chairman ROHRABACHER. Thirty seconds.

Mr. TUMLINSON. Thirty seconds? Well, frankly—you mean 30 seconds, really. Oh, dear. Anyway, I think we have a lot of misconceptions going on here. The fact is, the reason Sojourner was so small was there is no money, all right? The reason there is no money is that it's been eaten up by the Shuttle and the Station, and we have got to get ourselves to a situation where that money can be freed up so that we can do so much more.

People are impressed by what they are seeing, but it's what they are not seeing that's exciting. And we've got to be able to transfer those funds out of the Station and the Shuttle. I wish there were a Congressman here on this Committee from Missouri. He would be sitting there just simply going, show me, show me, show me, show me, show me, because we have got to see evidence of this.

We talk about privatization of the Shuttle. The fact is that the U.S.A. processing that's going on has, from what I understand, pieces of that processing have already been taken back by NASA. From what I understand, U.S.A. did make an effort at the direction of some folks in NASA, to try and privatize one Shuttle. They took that idea into NASA. They were cut off at the knees from what I understand. A couple of executives at U.S.A. had their heads rolled for even suggesting that they try and privatize the Shuttle. So again: show me, show me, show me.

The Station problem regarding the Russians——

Chairman ROHRABACHER. There's your 30 seconds.

Mr. TUMLINSON. Okay, then we won't go there.

Chairman ROHRABACHER. But just out of fairness, because whenever somebody from an agency comes before a Congressional Committee, everybody is always beating up on that agency. So I always try to make sure that even in Democrat Administrations, that we give the agency head the last 30 seconds.

Dan, would you like to have the last summary here, except for the Chairman's and the Ranking Member's summary, of course.

Mr. GOLDIN. NASA is proving that money is not the magic ingredient. The NASA budget doubled from 1983 to 1992. We hardly started any new programs. The key ingredient is not more money. The key ingredient is a focus on excellence and a vision of knowing where you are going and a commitment to the people that you have to get there.

The NASA budget has come down constantly since 1994, not in funny terms, but in real terms. And when you include inflation, it's even more. I think the record that NASA has stands for itself, and once again, I want to say I am so proud of team America. Thank you very much.

Chairman ROHRABACHER. Thank you very much. Mr. Gordon, would you like a few moments?

Mr. GORDON. Just very quickly, I want to take us back to the purpose of this meeting today, and that is to celebrate the 40th anniversary of NASA. I want to thank Director Goldin and his predecessors, the men and women that have worked for NASA, the contractors, really that have made us the envy of the world.

We are here all very proud of what you've done. You've inspired school kids and mothers and fathers all across this country and we are very grateful for the 40 years that you have contributed. And when history is written about those 40 years, there won't be a chapter, I don't think, that will any brighter. Thank you.

Chairman ROHRABACHER. And finally, let me just thank all of the witnesses for participating today, and Rosetel for demonstrating the new technology which brought us Pete Conrad all the way from Huntington Beach to the witness table.

I, too, am very proud of NASA. I am proud that we are a country where we have people like you, Dan, that are willing to sacrifice the number of years that you have; and all of you for the time and effort that you put into this hearing, but also for what you've done for our country. And Rick, for being the gadfly and the guy who challenges people to do it better and finds flaws in order to correct them.

That's what America is all about, and we believe that with technology and freedom, there's no limit to the potential of humankind. Today, we have been looking in the past and we've been face to face with the future, thanks to this technological demonstration. And I appreciate all of you and wish NASA a very happy 40th birthday and 40 more years of success.

This hearing is adjourned.

[Whereupon, at 2:04 p.m., the hearing was adjourned.]

○